Veganism, Archives, and Animals

This book explores the growing significance of veganism. It brings together important theoretical and empirical insights to offer a historical and contemporary analysis of veganism and our future co-existence with other animals.

Bringing together key concepts from geography, critical animal studies, and feminist theory this book critically addresses veganism as both a subject of study and a spatial approach to the self, society, and everyday life. The book draws upon empirical research through archival research, interviews with vegans in Britain, and a multispecies ethnography with chickens. It argues that the field of 'beyond-human geographies' needs to more seriously take into account veganism as a rising socio-political force and in academic theory. This book provides a unique and timely contribution to debates within animal studies and more-than-human geographies, providing novel insights into the complexities of caring beyond the human.

This book will appeal to students and scholars interested in geography, sociology, animal studies, food studies and consumption, and those researching veganism.

Catherine Oliver is a postdoctoral researcher, currently working on the ERC-funded *Urban Ecologies* project at the University of Cambridge. She completed her PhD on vegan geographies in 2020. Her research interests are veganism, beyond-human geographies, and friendship. Catherine can be found on twitter at @katiecmoliver.

Veganism, Archives, and Animals

Geographies of a Multispecies World

Catherine Oliver

Routledge
Taylor & Francis Group

LONDON AND NEW YORK

First published 2022
by Routledge
2 Park Square, Milton Park, Abingdon, Oxon OX14 4RN

and by Routledge
605 Third Avenue, New York, NY 10158

Routledge is an imprint of the Taylor & Francis Group, an informa business

British Library Cataloguing-in-Publication Data
A catalogue record for this book is available from the British Library

Library of Congress Cataloging-in-Publication Data
A catalog record has been requested for this book

ISBN: 978-0-367-69277-3 (hbk)
ISBN: 978-0-367-69278-0 (pbk)
ISBN: 978-1-003-14121-1 (ebk)

Typeset in Times New Roman
by KnowledgeWorks Global Ltd.

For Susan.

Contents

Figures

Preface and acknowledgements

On a dark winter night in November 2013, on the floor of my bedroom in the final year of my undergraduate degree, I watched the documentary 'Earthlings' (Monson, 2005). This film has a notoriety amongst vegans, sometimes referred to as 'the vegan-maker,' because of the disturbing and violent footage it contains. The film consists largely of undercover footage of the exploitation, lives, and deaths of animals in puppy mills, factory farms, research labs, entertainment animals, and the fashion industry. I cried. Knowing that through my choices, my body was fuelled by these wrongs, complicit in worlds of pain. I couldn't reconcile this knowledge with who I believed I was. My body rejected this new knowledge in the most obvious and visceral outpouring of disgust. I vomited. Up until this point in my life, I had been a vegetarian. The next morning, I went to the cupboards and fridge in my student house-share and removed everything that contained dairy or eggs and offered them to my housemates, telling them I had decided to become vegan. For months after, at random times, I would feel waves of horror and disgust come over me, leading to many nights of alternately tears and anger at the pain I could now see everywhere I turned. This transition to veganism was riddled with guilt, confusion, and a lot of strange meals with the limited availability of accessible and affordable vegan foods even just a few years ago.

Fairly soon after making this decision, I enrolled on Dr. Pat Noxolo's (who later became my PhD supervisor) undergraduate course on 'mediated geographies.' It was in this class that I began studying what I would later discover to be a long and rich history of geographical and sociological research on animals, interspecies relationships, and veganism. A couple of years later, in 2015, I began my doctoral research at the University of Birmingham. Over the course of the next four years, I had the pleasure of meeting not only many brilliant scholars, but also of interviewing and learning about the lives of vegans, past and present, across Britain. Knowing very little of the world of higher education, its demands, and rules, I slowly pieced my way through the first year of my PhD with something approaching a framework for a research project on contemporary veganism, specifically in the

West Midlands of England. However, two things happened that expanded and transformed this project. The first was taking a position on The British Library's PhD placement scheme in 2016, working in the archives of animal activist Richard D. Ryder. The second was in 2018, with the intervention by my mum in the lives of six chickens who were on their way to a battery egg farm to her house in Lancashire. And so, between the unique insight into the histories of animal activism through The British Library's archives, and developing a new mode of interspecies living with 'rehomed' chickens, the research and stories in this book emerged.

Britain has long declared itself a 'nation of animal lovers;' evidence for this is claimed at least as far back as Puritan rule under Oliver Cromwell in the 17th Century, when bloodsports became outlawed, and such practices became associated with the lower (and thus less 'civilised') classes. National animal loving has also been attributed to the Victorian era, with the rise of widespread pet-keeping, prior to which 'pets were often seen as an elite extravagance, and small dogs frequently appeared in satirical prints of aristocratic ladies, symbolising frivolity and indulgence' (Hamlett, 2019). The rise of dogs and cats as both domestic workers and companions drove the popularity of pets in Britain. The founding of the world's oldest animal protection society, the RSPCA, is also often used to qualify Britain's animal-loving history but, as Joe Wills writes, 'the frequent claim that Britain is a 'nation of animal lovers' can be hard to reconcile with the reality of how we often treat even the most revered of our fellow creatures' (2018, 407). This 'love' extends only to the animals we live with, not those who we use for food, entertainment, and medicine.

In his book, *Diet for a Large Planet* (2020), Chris Otter explores the history of humans, animals, plants, ecologies, and movement that powered Britain's industrial revolution. Notably, the 'nutrition transition' and 'meatification' of the British diet in the 19th Century led to a 'worlding' of the British diet, with the British diet's production moving outside of the national borders and implicating, via colonial rule, great swathes of the world as British agricultural hinterlands. Alongside this meatification and distancing of Britain from the production of animal agriculture was the early roots of a burgeoning alternative way of eating and living: vegetarianism. In September 1857, *The Vegetarian Society* was formed by a group of social reformers and devout Christians in Manchester, performing a contestation in and of a society where eating animals was on the increase, both through colonial expansion and as a symbol of wealth and upward social mobility (Gregory, 2007). Early vegetarians tried and failed to win the support of the conservative RSPCA, which is a theme common to the Society through the 20th Century, who determinedly focus only on domestic and working animals (as I discuss in Chapter 2). This early vegetarianism was, for many, deeply entangled with religion, especially Christianity, as well as other social reform causes such as poverty alleviation, food reform, and even socialism. However, it was not until 1944 with the founding of The Vegan Society that vegetarianism was formally split into those who ate dairy and eggs, and those who did not.

After the founding of the Vegan Society alongside Dorothy Watson, Donald Watson's writing reveals how the spiritual elements of eating persisted in his early British veganism: 'Even though the scientific evidence may be lacking, we shrewdly suspect that the great impediment to man's moral development may be that he is a parasite of lower forms of animal life. Investigation into the non-material (vibrational) properties of foods has barely begun, and it is not likely that the usual materialistic methods of research will be able to help much. But is it not possible that as a result of eliminating all animal vibrations from our diet we may discover the way to really healthy cell construction [and] a degree of intuition and psychic awareness unknown at present? (Watson, 1944).' A few years later, in the same publication, we also find the origins of the environmental roots of veganism not only to soil, but to national security and self-sustenance: 'the question of growing health foods is of real national importance, for no nation can be well which ignores the cultivation of its soil. We are taking a long time to learn that although we have a most fertile soil, we are practically a landless people' (Semple, 1945, 8).

Douglas Semple was an early member of the UK's Vegan Society. His concern with land use and urbanisation is contextualised at the time in concerns for self-sustainability in Britain being at odds with milk production, arguing that 'so long as we use dairy products we cannot make the most use of the land' (1945, 8), and also with in a long colonial exporting of food production (Otter, 2020). The echoes of this concern can be found today with regards to contemporary cattle farming, deforestation, and environmental damage. The connection between soil fertility, horticulture, and gardening remained central to The Vegan Society's communications and practices throughout the mid-20th Century, particularly amidst concerns for the future of feeding the earth's population in a time of land degradation (Henderson, 1948, 6–7; Smith, 1949, 7–8): 'To grow good crops it is essential that we co-operate with Nature and try to understand the relationship between soil fertility and healthy plant tissues' and 'if we would go to the root of our social and health problems we must individually live simpler and more natural lives' (Semple, 1947, 9–10). Animal activism, vegetarianism, and veganism have long been entangled with British society, not as an alternative way of living, but as deeply ingrained within and responding to wider social, cultural, and political issues.

In this book, I pick up the threads of activism and veganism across both historical and contemporary landscapes of veganism in Britain, where contemporary veganism has its roots as a political and ethical movement dating back at least to the 19th Century (Kean, 1998), and veganism and animal activism have been deeply entangled with other social reform and justice movements (see Oliver, 2018, where I have written about some pioneering women working across animals', women's, and LGBTQ+ issues). In thinking and researching across this elongated temporality, I understand how friendship has been and continues to be a vital source of strength for the organisation of activists, but also how they circulate around particular ideas of 'truth' to inform and share their practices. This is then troubled and tested by the introduction of six chickens into my life – Lacey, Bluebell, Olive, Cleo, Winnie, and Primrose – in 2018.

Where archival ethnographies and interviews offer insight into the practices, past and present, of those who advocate for animals, and to the lives of archival and abstract animals, it was only once I met these chickens that I realised the necessity to attend to their lives, and explore what veganism might mean, if anything, for them. The last two empirical chapters of this book are as such dedicated to the often-posed question of how we might live with other animals in a vegan world. In an ethnography with chickens (and other animals), I attend to how deliberate cultivations of space and interspecies relationships must be ingrained in vegan practices that are future-oriented. This is, of course, entangled with a suite of other social, cultural and political issues such as who has the space (and land ownership) to live with animals; who has the knowledge and time to adapt their lives; and how this is possible on a mass scale. I focus particularly on how this chicken-human relationship changes the space in which we live together as a multi-scalar transformation that intervenes into the agro-industrialisation of chickens (Davis, 2009). In 2020, with the Covid-19 lockdown in Britain, domestic chicken-keeping incurred a huge surge in interest, with the British Hen Welfare Trust – who rehome 60,000 laying hens annually from slaughter via regional centres – having received unprecedented numbers of requests to rehome hens (Oliver, 2020). Ex-laying hens being rehomed has the potential effect of disrupting the usual spatial separations between human and animal spaces to rethink who 'belongs' where. The lived reality of animals such as hens, their needs and welfare, as well as the hidden geographies of their exploited labour, are too often ignored even in the most radical imaginings of future spaces. Covid-19 has opened a series of broader ethical and practical questions around our relationships with other species, and brought home the importance of the critical political, ethical, and environmental challenges I attend to in this book.

This book, and the research contained within it, would not have been possible without the work of generations of tireless advocates and thinkers. I am grateful to all those who have participated in revealing and resisting the abuses of animals, and in rethinking how it is possible to live with animals. In particular, this work owes a huge debt to the thinking and practice of many activists and scholars (and activist-scholars), but in particular I am grateful for the kindness and work of Kim Stallwood and Carol J. Adams. In addition, I am especially grateful to Richard D. Ryder for preserving his important archive at The British Library, allowing me to work within these histories, and to The British Library, Polly Russell, Gill Ridgeley, and Jonathan Pledge for homing me via their PhD Placement scheme in 2016–2017.

Similarly, I would like to thank all the advocates and activists who, over many years, have contributed to the progress and growth not only of veganism, but of reimagining the contours of multispecies life and embodying the politics and practice of friendship I write about in these pages. Especially, I am grateful to the vegans who spoke with me in interviews, and those who I had the pleasure of working with during and beyond the scope of this research. I am also grateful to the whole team at Routledge for publishing this book, especially to Faye Leerink for her kind and constructive feedback, to Nonita Saha for her help and editorial guidance, my copyeditor Samynathan Mani, to Richard Twine for his kind feedback, and to the other anonymous reviewers for their thoughtful comments and engagements with this book.

Writing this book was made possible only through the guidance and mentorship of my PhD supervisor, Pat Noxolo, whose unwavering support for my work has allowed me the space to flourish and develop my thinking and research. Additionally, I am grateful to the examiners of my PhD, Peter Kraftl and Eva Giraud, for their generous and ongoing engagement with my research. During the course of the research and writing of this book, I was also incredibly fortunate to have the friendship of many incredible people. In particular, I am grateful to all of my PhD colleagues, academic mentors, and friends in the School of Geography and beyond at the University of Birmingham, especially to Milly. I am also thankful to the Urban Ecologies team and colleagues in the School of Geography at the University of Cambridge for their challenging and thoughtful engagement with my work. My research is also indebted to my wonderful friends, particularly Andy and Lisa, who I was fortunate enough to meet and work with at The Vegan Grindhouse. Although I won't name each individually, my friends across the world, both inside and outside of the academy, have all been a joy to know and a huge source of support and strength throughout this project. My special thanks go to my closest friends, Phil Emmerson and Faye Shortland, for everything, and to Catherine, Lizzie, Chloe, Laura, and Megan who have heard more than any friends should need to about vegan geographies.

My work has always been a collective endeavour, and I hope that this is reflected in this book. This collectivity has never been only human but is situated within beyond-human collectives, without whom my life would have been less interesting. I am particularly thankful to Lacey, Bluebell, Olive, Cleo, Winnie, and Primrose, and to Charlie and Fizz for their friendship whilst working on this project.

Finally, thank you to my partner Liam who has carefully and tirelessly supported me as I researched and wrote this book, and to my mum Susan, to whom this book is dedicated, for being my guiding light.

Introduction

I became vegan in November 2013 after watching the documentary 'Earthlings' (Monson, 2005). When I started writing this book, I decided to watch at least the beginning of the film again wondering if, years later, I would react similarly, having viewed similar images to those within the film regularly since. For some vegans, watching mediated animal lives and deaths strengthens their beliefs and is an important factor in practicing veganism and driving activism. This imagery can offer possibilities and impossibilities for the transformation of food practices and systems (Goodman, 2018). For other vegans, witnessing pain and violence can lead to experiencing symptoms of compassion fatigue, or 'secondary traumatic stress disorder' (Figley, 2013), often found in those treating the traumatised, as a 'cost of caring' (ibid., 1). Joy applies this idea towards those acting for animals as carnism-induced trauma: as witnesses we are also victims (Joy, 2009). In opening to the 'truths' of veganism (where the language of 'truth' is taken directly from the vegans I interviewed in Part Two), vegans also open themselves to a transformed way of navigating the world, at once disturbed and disturbing. Opening the video for Earthlings, I feel the tears rise up: 'earth·ling *n*: One who inhabits the earth' (Monson, 2005).

I managed to watch eight minutes of the film. Even eight years later, I cannot write of my own becoming vegan eloquently, because it is an embodied and emotional knowledge that I carry with me. It clashes with the world I inhabit every day, where 'for many of us who take on that knowledge, it is a strange, bifurcated existence. I live in my vegan world, in which I see what I think is food; and I live in a meat-eating universe' (Stallwood, 2013, 44). Living in a heightened awareness of non-human suffering is spatially disturbing. When walking from his home to the train station, Richard White writes that were he 'to push an observer to move beyond an anthropocentric scripting of this encounter with place and ask that they critically focus instead on the excessively obvious presence (or indeed absence) of more than human animals' (2015, 213), they would find a disturbing urban narrative, rendering violence so commonplace it is invisible in its omnipresence.

When paying attention to concealed interspecies violence, pet shops, fast-food restaurants and butchers become part of this disturbed and disturbing narrative of place, and sense of the world that cannot be unknown (McDonald, 2000). Sadness, anger and despair transform the world and hope, action and transformation seem so distant. These latter motivations and feelings emerge, for most vegans, later in their vegan transition (explored in Chapter 5). The closeness of this work, and my own veganism, mean the separation between field and not-field is not distinct, but entangled and destabilised (Hyndman, 2001), 'inverting assumptions about home and field... and the taken-forgrantedness that the field is always somewhere else' (265). A relational and unbounded field is not static and elsewhere but carried with me, shaping my knowing and being in the world within and beyond the spatio-temporal confines of this research.

Veganism is generative in grasping how less violent multispecies worlds can be imagined and enacted. In this book, veganism's power and potential for the future is demonstrated through empirical accounts of historical and contemporary activism rooted in feminist care and extending friendship beyond the human. Throughout the theoretical and empirical work, this book explores the emergence and growth of vegan beliefs and practices, and how these have and continue to transform space and society in Britain. In this book, I contend that animals and our relationships with them are always part of an ethically and politically informed and motivated 'beyond-human' geography. This term draws attention to anti-hierarchical multispecies relationships and goes further than recent accounts of entanglement to destabilise the centrality of the human in human geography. The development of this 'beyond' space that elongates space and time is in contention with and resisting the romanticising (in the more-than-human) and/or subjugating (in the non-human) of animals as mutual constitutors of space, time and the world.

In this book, I contend that the geographical study of veganism pushes 'more-than-human' geographies and critical animal studies to take seriously alternative interspecies relationships in the face of uncertain worldly futures. No longer a 'fringe' belief and practice, veganism has firmly entered the frame. The work in this book is located in Britain but overlaps with a larger European/American/Australian vegan movement, due to the archives and location of my interviews. Veganism is not, however, a Western movement. There are indigenous and majority world practices of plant-based eating that far predate the histories in this book (see Zuri, 2021; Polish, 2016; Harper, 2011; and Ko and Ko, 2017).

Veganism's nurture and care for other species and the environment has a history of being positioned as *women's work* (Adams and Gruen, 2014). Contemporary veganism has enjoyed growth and visibility that is in part, although not entirely, linked to a particular kind of femininity through the rise of Instagram's wellness aesthetic of thin, young, white and wealthy women (see Greenebaum and Dexter, 2018). The landscape of veganism has transformed quickly in Britain, with increasing visibility on supermarket shelves (being Britain's fastest growing food product market,

LiveKindly, 2020); the first vegan cooking show, Matt Pritchard's *Dirty Vegan*, launched on the BBC in 2019; and The Economist labelling 2019 '*The Year of The Vegan.*' This growing visibility has put veganism under scrutiny, moving from an extreme fringe belief into the cultural mainstream. But, instead of a growing diversity of leaders and figures, veganism is continually represented by the same kinds of white men who claim their presence is legitimising of an otherwise feminised, vegetal, and therefore dismissed movement. This is a historical pattern of marginalising and sentimentalising this women's work, as I discuss in Chapter 2.

In independent surveys commissioned by The Vegan Society (2019), practicing vegans have been reported as quadrupling in the UK between 2014 and 2019 (150,000 or 0.25% of the population, to 600,000 or 1.16%). Those most commonly following a vegan diet were women aged between 18 and 34 years, at 3% of this population (Statista, 2019). In their most recent data, The Vegan Society (2016) found that twice as many women as men were vegan, which is echoed in statistics from the USA (where up to 6% of the population identified as vegan in 2020). For surveyed vegetarians, there was a split of 59% women to 41% men divide, but for veganism this shot to 79% women (Humane Research Council, 2014). Because of the recent growth in veganism, some of these numbers may already be unrepresentative, and the lack of gender options skews these numbers. However, even as veganism grows, it is not at the rates expected by recent surges in plant-based products flooding the UK market which points towards an increased *visibility* of veganism within an increase in plant-based eating by flexitarians, vegetarians, and omnivores alike.

In the remainder of this introduction, I deal directly with four matters critical to establishing the assumptions and knowledges upon which this book rests. First, I briefly introduce and provide an overview of the contemporary debates in animal geographies, critical animal geographies, and vegan geographies that guide the work in this book. The second matter relates to the definition of veganism and its relation to other forms of animal activism that are in this book taken as distinctly (and problematically) "British" modalities, a notion that is troubled in Chapter 2. Thirdly, I introduce three key guiding themes of the book that loosely align with the empirical work which the majority of the book is dedicated to. These are *friendship*, which is central to the archival work of Chapters 3 and 4; vegan constructions of '*truth*,' which guide the discussion of the contemporary movement in Britain, drawing on activist interviews, in Chapters 5 and 6; and *temporality*, which is particularly picked up in Chapters 7 and 8 in multispecies ethnographies with chickens. However, it is important to state that these are not clearly delineated or separated categories, and each of these crosses into all seven chapters to present an interwoven meshwork across time, space and species. Finally, I include a methodological summary of my empirical work in the Ryder archives, interviewing vegans, and living with chickens, which is further developed and shared within the empirical chapters themselves. Finally, I reflect briefly on modes of writing beyond the human and offer a brief summary of each of the chapters.

Vegan geographies

A contemporary geographical concern with veganism, and what this means for animals and multispecies space, is situated within a long geographical disciplinary interest in animals, dating back to Newbigin's 1913 book *Animal Geography* (Wolch et al., 2003). This interest in animals has progressed differently across animal geographies, critical animal geographies and emerging vegan geographies. Animal geography situates animals within assemblages of things and places constituting the multinatural (Lorimer, 2012). Despite its concerns with places, processes and the ordering of society and environment, animal geography traditionally has not engaged with critiquing the status or human/non-human relations (Castree, 2000). Critical animal geography is, conversely, concerned with the problems human exceptionalism poses, where animals are understood as 'subjects of and in spatially uneven practices' (Hobson, 2007, 253). Where animal geography sometimes lacks a grounding within longer animal studies philosophies and politics, critical animal geographies contest this human-animal border and the exclusions and inclusions of always already multispecies spaces (Wolch and Emel, 1995; Philo and Wilbert, 2000; Wolch et al., 2003), critiquing the relationships between animals and space where the 'fundamental interconnectedness of humans and animals … authorizes oppressions based on gender is the same ideology which sanctions the oppression of animals' (Hovorka, 2015, 5). This work is largely undertaken in political studies, sociology, philosophy and history and the interdisciplinary field of (critical) animal studies.

But, as The Vegan Geographies Collective (Hodge et al., 2017, np) call attention to, there remains a 'scarcity of available literature [which] highlights the need for geographers to further reflect on vegan activism and practice.' While work in geography on and with animals has long been a mainstay of the discipline, humans' changing relationships with animals in society, culture and politics through veganism has received less explicit attention, especially given its rising visibility and popularity in contemporary Britain especially. For geographers, rethinking human relationships with animals does not take place in a theoretical abstraction, but rather demands that we attend to questions such as 'how can we more justly share space?' (Collard and Gillespie, 2015, 8). Geographers thus have a critical role to understanding veganism as an inherently spatial praxis that has yet to fully emerge in the discipline.

The work in this book is borne out of cross-disciplinary traditions and critical knowledges, ultimately arguing for a geography that further politicises and critiques the relationships between animals and society, seeking to imagine and practice a somehow different emancipatory multispecies future, and to follow geographical work that troubles the boundaries of the human and of 'human' geography (Srinivasan, 2015). As such, this book follows in the call that there remains a need for a geography that matters and for a geography that makes a difference (Massey et al., 1999) to be expanded, theorised, and practiced beyond the human.

Animal welfare, rights, and veganism

Veganism is related to both animal welfare and animal rights movements in Britain, where these specific forms of Anglo-American animal activism underpin contemporary Western veganism. To locate this book, a short overview of these histories and definitions of animal welfare and rights activism and their ongoing influence on and presence within veganism is included here.

Animal welfare activism calls for incremental change to reduce the suffering of animals and does not necessarily require a commitment to ending animals' use for and by humans. Welfarism calls for the conditions of the animal-prisoner to be improved: bigger cages for chickens, slaughter before transportation and restrictions on animal testing. Where veganism is concerned with individual, collective and worldly consequences of the agricultural-industrial complex, animal welfare activism is regularly called out on its perceived hypocrisy. Animal welfare causes largely appeal to the general (non-vegan) population to feel that the animals they consume lived a good life, but remain killable (Morin, 2018). Despite critiques of its moral inconsistency, incrementalism is still present within veganism, as a long-term approach to the eradication of [ab]using animals (Francione, 2010). As such, some vegan activism focuses on intervening and improving animal welfare with farmers, animal testers, politicians and other parties to move towards alternatives. Appeals to animal welfare are also contemporary strategies within veganism rather than solely part of the dominant independent movement that they were throughout the 19th and first half of the 20th Century (Kean, 1998).

With the advent of the *animal rights movement* in the mid-20th Century, the inclusion and participation of animals in society and politics through discussions about animals' rights and human's duties were mainstreamed (Regan, 1983). Animal rights activism includes a broad spectrum of positions across those who believe in rights for some animals, some rights for all animals, and often in the maintenance of a 'rights' hierarchy with humans at the top. As such, within some animal rights accounts, rights for animals are not mutually exclusive with eating them, because rights are contingent and 'given' by the human, not necessarily ending hierarchy but possibly reinforcing it (Giraud, 2013). Intellectual manoeuvres (Brophy, 1971), the limits of imagination (Godlovitch, 1971) and a politics of doubt (Wadiwel, 2016) are deployed within these variously sentientist politics (Cochrane, 2018) to dictate who matters enough for conditional inclusion in anthropocentric rights-based frameworks. In the latter 20th Century, the rise of rights-based activism was associated with liberationist tactics and a 'no compromise' stance.

With the advent of animal rights activism came a masculinisation of animal activism, which removed care and sentimentality, seeking to rationalise empathetic beyond-human relationships as only a matter of righteousness and justice, rooted not in love but in truth. This rise of rationality, typified by Singer (1975), side-lined feminised and (eco)feminist work of care and love (Adams, 1994) with the aim of legitimising animal activism through masculinist discourse and action (Fraiman, 2012). Since the 1990's, animal activists have diversified their agenda, maintaining both welfarist and rights-based activism but also aspiring to professionalisation (Wrenn, 2012). Animal welfare and rights activist principles continue to inform veganism and are often deployed as an agenda of veganism with little critique of their anthropocentric frames. Veganism's contemporary prevalence is a continuation and separation from welfarist and rights activism pasts, which takes seriously a commitment to anti-speciesism, a rejection of which is central to the practice of and belief in ethico-political veganism alongside environmental and health-based concerns. Animal welfarism and animal rights did not develop in a linear progression towards veganism but continue to exist within veganism, as well as independently of it.

In 1981, Brigid Brophy gave the inaugural address at the Council for the Prevention of Angling in London. When asked to rank the cruelties done to animals – a question familiar to all of us who work with and for animals – Brophy replied: 'The question is not designed to be answered. It is designed to provoke an interesting and cosy intellectual discussion, in the course of which the questioner can stifle the promptings he feels from his own conscience towards doing something about any of the atrocities he names; and it is designed to divide the now quite large and certainly growing pro animals-in-general movement' (Brophy, cited in Stallwood, 2015, 7). Kim Stallwood (2013) has argued that ethical and political concerns for animals began to constitute a cohesive movement with Brigid Brophy's article 16 years earlier, *The Rights of Animals* (1965). This moment of creative construction was part of a turn in animal activism (Stallwood, 2013) that did not begin with but, I argue, informs contemporary veganism.

Brigid Brophy situated human-animal relationships as 'unremitting exploitation,' whether sacrificial, machinic or of consumption, following Harrison's *Animal Machines* (1964) to critique animal abuse in all its forms, intensities and directions. Since Brophy, animals' mattering has been continually debated. This book, like much contemporary animal work, refuses this debate (Hadley, 2015) and is instead concerned with the ethical, political and geographical consequences of the fact that animals *already* matter.

Friendship, truth, & temporality

In this book, I take up the interrelated questions of friendship, truth, and temporality to approach and understand veganism and beyond-human geographies. I briefly introduce these concepts and their application here.

Where heteroromantic and familial structures are upheld as primary dependencies (hooks, 2000), the role of **friendship** in making and sustaining politics, culture and society is largely seconded to these other kinds of relationships. In activism and leftist politics, comradeship in particular is symbolic of politicised relationships: a performative naming to signal a common cause, usually hyper-masculinised. This centring of comradeship over friendship in animal activism is related to masculinisation and 'pussy panic' of activists and scholars (Fraiman, 2012), whose closeness to feminised subjects (animals) simultaneously affords them privilege in becoming leaders in these areas whilst threatening their masculine identity. Comradery has less accountability and bound within a cause, ensuring masculinity can remain intact. Friendship is centred here instead, as a resistance to the masculinisation of the subject and as part of elucidating a different kind of politics, informed by feminist practices.

Friendship has been conceptualised by white, male, Western thinkers as 'so tightly linked to the definition of philosophy... that without it, philosophy would not really be possible' (Agamben, 2009). This deployment is an active relation of power that dictates who is included and excluded as possible thinkers, by their potential as friends (Deleuze and Guattari, 1994). This mobilisation of friendship in Western imperialist philosophy is, as I discuss throughout this book, a protectionist misnaming of friendship that upholds hierarchy. This misappropriation of friendship by the powerful reifies it as elitist and exclusionary, rather than an open (yet bounded) network of relations of care and affection, imbued with ethical and political intention and direction. When friendship is an approach – a way of life – it instead yields a particular culture and ethics outside of exploitative hierarchy (Foucault, 1997, 137).

Friendship as a way of life for those on the margins goes beyond cultures and ethics of communities in an approach and praxis that shapes action, activism and space. In veganism, friendship between humans must additionally be able to be extended to friendships between humans and animals, where all parties are understood to be engaged in (real or imagined) ethico-political relationships. This requires that we know animals well enough (Midgley, 1983) in both proximity and difference to maintain their inclusion within such an expansive relation. Which is to say, friendship is a feminist ethico-political approach and practice of freedom in the relations between humans and humans, humans and animals, animals and animals. Through friendship, those already marginalised protect, organise and resist on behalf of the friend. Friendship is a political, social and cultural geographical concept that I explore, notably in Chapters 2 and 3, as potentially transformative to both interspecies solidarities and the establishment of alternatives modes of living with and advocating for other animals.

This permeation and transformation of everyday life in veganism is intimately connected to the ways in which veganism and animal activism circulates around particular '**truths**.' This is not concerned with demanding a singular 'truth,' but rather in revealing and understanding how and why vegans circulate and collectivise around fluid and relational 'truths.' Veganism has always been concerned with finding and exposing the 'truth' of humans' treatment of animals. Through undercover exposés, literature and protest, vegans who have been exposed to the 'truth' of these systems often feel it is their duty to share and educate others. What is particularly revealed through the interviews in this book are the socio-spatial consequences of this 'truth' on individuals, communities and the world within and outwith veganism itself. The multidirectional echoes of learning about and entering the world of (not) eating are navigated by vegans in painful disruptions of all they thought they knew; they are confronted continually with those denying their 'truth.' As such, the notion of 'truth' is within this book explored as personal, partial and embodied and, in contemporary veganism, this embodied truth is being foregrounded as a way of knowing and experiencing the world as a vegan. The use of 'truth' is as a contested, fluid, and unstable concept. Furthermore, the use of 'truth' is rooted in the empirical research, drawing from the interviews with vegans and activists (Chapters 4 and 5). This is not a move to claim a particular 'truth,' but rather to explore the spatial and embodied ways 'truth' is realised in veganism.

The contemporary vegan sensibility of 'truth' explored in this book is not, as in historical animal activism, centred around utilitarian or rational arguments of rights and suffering (Chapter 2). Rather, it is deeply entangled with an *embodied* sense of wrongness in eating animals. This feeling of discomfort and wrongness is realised through encounters not only with actual animals, eliciting interspecies empathy, but also through virtually mediated spaces of films and social media that revealed the pain and violence of eating animals, as in my own Earthlings encounter. The embodiment of veganism as an ultimate 'truth' affects navigations of space in subtle and explicit ways. Ordinary spaces become the holder for extraordinary transformative events where vegan transitions occur; the exceptional is revealed through the exposure of violence underneath or behind the everyday (White, 2015). For vegans, their transition is often positioned as one of the pivotal moments of their lives, fundamentally refiguring the ways in which they relate not only to themselves and to other people, but to society and space itself. With this in mind, veganism's growth is an important site and force for geographical understandings and theorisations of the contemporary world through eating and activism.

Veganism creates a 'truth' that rubs against others' worldviews, but this truth is not singular, fixed nor agreed upon. The claims to a 'truth' is regularly disturbed and challenged within veganism from definitions to practices, revealing the malleability, fluidity, and contested nature of veganism itself as socio-spatially and historically contingent. Moving towards the possibility of vegan techno-futures has seen these debates expand into new frontiers of, for example, cultured meats and capitalist veganism. Policing the boundaries of veganism often leads to fraught clashes of opinion within vegan communities, with disagreements over infractions of the principles and practices of veganism flaring up both online and offline. This book does not attempt to offer solid 'truths' to one or other side of contemporary vegan debates. Rather, I seek to understand how veganism's truths define its community and distances members from former relationships, disrupting and making strange people, society and the world. To do so, I share the stories of vegans in Britain, their relationships with 'truth' before, during, and after veganism in their everyday lives and spaces. This 'truth' is not only an orienting principle, but a relational network of actors, and a mode of expanding veganism through outreach as education. Attending to navigations of their relationships with themselves, their friends and family, and the world itself, veganism is proposed as a narrative for dealing with their own past violence, transforming themselves as a project of veganism, and a geographical practice of re-imagining the future.

As such, the work in this book is inescapably concerned with beyond-human **temporalities**, conversations which particularly come to fruition on enacting possible futures in Chapters 6 and 7. This project's temporalities are explicated in the multi-sited geographies and methodologies addressing the historical geographies of animal activism; interviews with contemporary veganism; and in experimenting with potential imaginings of beyond-human futures through multispecies ethnography. The temporal elements of this book are thus, at times, explicitly attended to. Temporality is also attended to as a (dis-)organising mechanism of actors and events and the medium through which vegan networks and knowledges emerge. This temporality is, following Massey (2005, 30), 'thoroughly spatialized,' whereby 'if time presents us with the opportunities of change and the terror of death, then space presents us with the social in the widest sense: the challenge of our constitutive interrelatedness' (ibid., 195).

Across historical and contemporary animal activism and veganism, there is an inherent future-orientation, through which other modes of living with and without nonhumans are imagined and fought for, or against. Veganism is always in motion and interconnected through persistent and enduring relations of care, which I seek to represent in this book. Examples of these connected temporalities can be found in the notion of a historical idealised and radical activist as liberationist persists in contemporary veganism, making people less inclined to identify as an activist (Chapter 4); or the archival bond between the reader and the stories of animals can affect the present in determining and sharing genealogies of human domination, and animal resistance (Chapter 1). These beyond-human worlds are entwined in politicised and expansive spatio-temporal networks of care where relations shift within an open but finite history (Nancy, 1990). These temporalities bring history into the present, such as in the Case of the Brown Dog in Chapter 3 or the slowed time of interspecies togetherness in Chapter 7.

Temporality is particularly entangled with notions of 'the beyond' and the 'beyond-human,' where the beyond is something impenetrable between us, holding us both apart and together. The beyond is a site for critical inquiry and intervention between and within multispecies worlds, differently sensing and navigating temporal, spatial and bodily distances and other species' experiences of them. The beyond is spatio-temporally and socially elongated, allowing for the space between us to become one we might cultivate to allow for spontaneous encounters and longer-term commitments. Where interspecies trauma has a destructive relationship with temporality, the beyond can mess with linear temporalities (Povinelli, 2018). In Chapters 6 and 7, I ask how these traumas can be brought into the present and renegotiated when individuals of different species choose to live together in ways that undo historical and contemporary violence. This beyond-human geography asks questions of what the future might look like through transformative relations in the present.

Friendship, truth and temporality across space and species are disruptive. Within and beyond human collectives, they demand (re)navigations of proximal and somatic difference and distance. In this book, they are the medium through which, and without which it would be impossible, to explore if and how we might rethink 'us' to include other-than-human beings.

Methodologies

The fieldwork for this book began in September 2016 in the archives of Richard D. Ryder at the British Library, where I spent two or three days a week for eight months. As part of their PhD placement scheme, my time was spent working in their basements with the uncatalogued archive. This archival period was followed by a reflective period working with the archival data and field notes. Then, almost coinciding with the end of my archival fieldwork, my mum rescued six battery hens from a farm just a few hundred meters from her home. Over the next months, I spent time with these chickens, marking multispecies ethnographies beginning in April 2017. Finally, interview recruitment took place between January and March 2018, and interviews took place between March and August 2018.

The Ryder Papers were donated in three parts to the British Library, the first arriving in 1999. This was donated as an animal welfare archive which documented the intellectual foundations and practical strategies of an important, and still growing, force within British pressure-group politics. The archive has around 70 boxes of materials, consisting of correspondence, press cuttings, campaign materials, memoranda, drafts of papers and books, audio and video tapes and photographs. The first deposit came off embargo in 2005, the second in 2015 and the third remains embargoed until 2025. These are largely dictated by dates of materials, but some selected sensitive materials remain embargoed. Although the first two loads are now no longer embargoed, they were, at the time of research, yet to be catalogued and only the first load had been sorted. Few people outside of the British Library institution have access to the archives in this state, unsupervised and in their totality. As somewhat part of the institution for eight months, although not fully, I developed a unique relationship with(in) this archive, here understood tentatively as a friendship towards and beyond the archive, as discussed in Chapter 3.

The next phase of my research was to interview vegans (Chapters 4 and 5). When I began researching veganism in 2014, it was already mainstreaming (Wrenn, 2012) and there has since been a surge in veganism, with 16% of new food product releases being vegan in 2019 (Young in the Independent, 2020). This raises questions, however, of who these products are for, when the actual numbers of people identifying as vegan, although increasing as noted earlier in this chapter, are nowhere near this high. Nonetheless, the timing of my interviews in 2018 meant that there was a relatively large pool of potential participants, but a small sample was selected for feminist semi-structured interviews. I selected sixteen interviewees who reflected a range of ages, occupations, genders and races. All of my participants were based in Britain and were aged between 21 and 70. The conversations were open and fluid and reflect the current moment of growth and trends in veganism. My interviews followed feminist approaches, whereby comfort and connection were prioritised (McDowell, 1997). The questions were designed to build conversations about veganism, before moving through pasts, presents and futures before returning finally to reflect on what veganism means and activism is to them, and how they embodied and practiced this within their own lives. This approach to the interviews created space for voices and experiences of the participants to be represented in storying as geographical method (Besio, 2005) within and across the spatial and temporal boundaries of this work.

Throughout the interview process, I found myself often thinking of a conversation from several years before. While working at a vegan burger business, we were cooking burgers and talking about my colleague's dog. Somehow, we got onto the subject of how rarely our veganism comes up here at work, where we are all vegan. We talk about recipes, meals and vegan news, but the questioning of our veganism is absent. It seems odd. I didn't even know why my colleague became vegan. It's almost as if there is an unspoken bond holding us together. We can almost ignore our veganism here in this vegan workplace. It's somehow different to 'elsewhere.' My colleague agreed with me, that it was nice come to work and relax.

I found this same sense of ease in the interviews. My conversational approach allowed participants to lead, identifying elements of the questions that they were more receptive to and shaping the conversation (Daigle, 2016). There was a felt bond that we occupied the same network with our values and ideals. The conversation often went off on tangents, bleeding into wider sharing about their lives, families and friends and even friends we had in common. For example, interviewing Shane, he circled back regularly to how when entering a space, knowing it was a vegan one changed that space into one of safety: 'there are really positive, life-affirming ways that veganism can impact your relationships with others who are known to you and others in your immediate circle and I try and make it all a positive experience, I think veganism is a positive thing, choosing compassion, wanting to be nice' (Shane, June 2018).

Hyndman (2001) conceives of field/work as transcending 'here-and-now' and 'there-and-then,' across time and space. However, I would move away from separating the here-and-now from the there-and-then, rather seeing these as mutually constitutive. The final part of this research, the ethnography, challenged me not only intellectually, but also to reflect on the ways I live with others and my own practices and beliefs, understanding how the historical there-and-then has shaped the present, but also how our own here-and-now is the location from which we imagine and enact future theres-and-thens. My ethnography took place alongside and after the other methods, encompassing distinct but connected 'encounters, seeking to understand 'how a collective movement [can] reinvent the definition of the subjective self' (Braidotti, 1994, 180). The somatophobia of research (Spelman, 1988) prioritises a disembodied and distanced analysis of the world, reaching for a view from outside, above (Haraway, 1988). This ethnography follows much contemporary feminist geographical scholarship in considering everyday life and the mundane and habitual lives of activists as of importance to their activism (Vaneigem, 1983). The ethnography in this book is informed by feminist theory. In it, I engage with 'shadowy regions of the mind where anger and rebellion about sociopolitical realities combine with the intense desire to achieve change' (Braidotti, 1994, 179).

Primarily, my ethnography took place with six chickens, who came to live with my parents in 2018, rehomed from a commercial farm (Chapters 6 and 7). The multispecies ethnography was not an intensive period but spaced over encounters and relationships over more than two years. More distant animals also feature in this multispecies ethnography (Kirksey and Helmreich, 2010): pigeons, foxes and herons are the most frequent friends, but so were worms, bees and geese. This multispecies ethnography explored the vitality and precarity of these animals and also the objectified, fragmented, and consumed animal-bodies (Adams, 2010) – or animal body-parts – who I encountered daily in supermarkets and restaurants. This ethnography is concerned with the precarious lives of animals (Stanescu, 2012) but also with the temporal dimensions of life through a concern for those animals disappearing from our everyday lives who constitute a 'multitude of different creative agents' (Hardt and Negri, 2004, 92) within the ethnographic field.

These spaces of ethnography were multiple, fluctuating and contradictory, imbued with meaning beyond the human: 'think of being in the middle of an ocean, of always being in its middle. Most humans know the ocean from its edge, standing on the liminal shore looking out. But from the middle we may envision the 'complex relation between different velocities, between deceleration and acceleration of particles' (Deleuze, cited in Probyn, 2016, 15). This middle is where this ethnography is located: the in-betweenness of everywhere, somewhere and nowhere.

Writing beyond the human

When we imagine the future, the human brain uses the same areas to remember past experiences and dream as to imagine future spaces, collectivities and events. When we imagine in the future, we are engaging in a historical dream that disrupts temporality. Writing, as a representation and account of life and events, is deployed to traverse spatiality, temporality and species, affected by the content and work itself. Following Pyke, I am 'attracted to an ethics of writing that works to move beyond this humanist paradigm' (2019, 4) of anthropomorphising animals. Reading animals as extensions of or supplementary to the human has been challenged by learning from narrative ethology (McHugh, 2011) and developing new collaborative learning frames (Bekoff, 2002). Beyond-human worlds often elude capture where writing alone cannot do justice to the affective, interwoven narratives of multispecies and elongated temporal practices of caring about animals. To conclude this introduction, I briefly outline the chapters of this book and how they contribute to the overall project of understanding veganism, animal activism, and multispecies worlds through friendship, truth, and temporality.

In Chapter 1, I illustrate the permeation of binary thinking in animal activist histories and critique this persistence of the human/animal. The work in this and the following two chapters draws on archival research undertaken in the archives of Richard D. Ryder. In this chapter, I grapple with the historical contingencies of interspecies care and activism. This builds to propose an embodied approach to the multispecies, which allows relations between and beyond the present to be rescaled, which echoes throughout the book.

Chapter 2 explores how the centring of white, upper/middle class male subjects (Adams, 1994) has permeated animal activism across intellectual, organisational and radical typologies, with white Western men being deemed legitimisers of feminised praxis of caring (Fraiman, 2012). This history of white, masculine rationality as a legitimiser of beyond-human care is never far from mind in the archives of Richard D. Ryder. In this chapter, questions of violence are foregrounded as I navigate the archive that stores animals' skin as book covers, within an institution itself steeped in colonial reproductions.

Following from this, Chapter 3 focuses on encounters with animals and humans of and in the archives, tracing particular stories of animal activism to construct a trajectory of friendship beyond the archive, positioning the enduring pulse of friendship as shaping contemporary networks of veganism in Britain. Historical friendships reconstitute understandings of vegan histories, as well as elucidating how these relationships and approaches of, to and with friendship are necessary to understanding ethico-political veganism between humans and extending this to animals.

In Chapter 4, I turn to contemporary veganism and explore how the mainstreaming of veganism has, as well as seeing the espousing of hostility and extremism of vegans, offered new ways of relating and understanding veganism as a *tripartite* practice for health, animals, and the environment. In exploring the contemporary landscape of vegan activism, this chapter explores the historical legacies and contingencies of animal activism, the spectrum of tactics, and the proliferation and successes of *quiet activism* approaches tentatively offered in interviews with vegans.

Accordingly, Chapter 5 follows from this, and connects variegated stories-so-far of humans and animals within elongated spatio-temporal networks through different temporal feelings of, performances of, recurring and disturbances of truth. In particular, I explore how embodied knowledges are entangled with vegan truths and how 'feeling-wrong' is explored as encompassing variously a sense of injustice, a disturbed sense of place and distressing (bodily) implications of pain and death in veganism, drawing on interviews.

Chapter 6 is the first of two chapters focussing on multispecies ethnographies with ex-farmed laying hens. In this chapter, I introduce Lacey, Cleo, Bluebell, Primrose, Olive and Winnie, as well as the space we shared together. In doing so, I think with and about those others who care beyond themselves about animals who are differently navigating the world, disturbing dominant truths and navigating interspecies relations through re-navigations of somatic and proximal distance and closeness in a continued trajectory of friendship through both veganism and activism.

In Chapter 7, I continue to work with multispecies ethnography, but this chapter is focussed more determinedly on bringing into conversation friendship, truth and temporality in this beyond-human geography with six chickens (and other animals). Living with chickens entails questioning personal, collective and worldly knowings and doings. When truth becomes untruth, uncertainty transforms the world into a strange, bifurcated one, disturbed and disturbing. The field disperses but remains in my imagining as it was, not as it is and, in this chapter, I attempt to theorise the space beyond and between us.

Finally, in the conclusion, I offer insights into how these strange, winged creatures, chickens, disrupted and disturbed my navigation of the world and my understandings of veganism, even after they died and the material relation between us has ended. Then, I trace how beyond-human geographies and care have been, are, and might be being practiced through veganism as multispecies caring through refusal, but also a building of an alternative geographical imagination and of multispecies worlds. Geographers thus have a critical role to understanding veganism as an inherently spatial praxis that has yet to fully emerge in the discipline, with expansive temporal and spatial implications and realisations that are growing in both visibility and importance in the world around us.

As Stauffer (2015, 7) writes 'when no one listens, stories get lost. But unaddressed harms do not disappear. They remain... haunting the official sites of transition and reconciliation.' The ability to hear, and thus write, is always affected by what is already known. If we do not want to know, then what we may hear and attend to is shaped by a desire to avoid this knowledge (McDonald, 2000). I intend in this book, through the archival research, interviews and multispecies ethnography to understand friendship, 'truth,' and temporality in vegan beyond-human worlds. In so doing, I understand how animal activists and vegans have historically and contemporaneously navigated spatial, temporal and species differences and distances, and how they continue to do so by imagining and establishing different futures.

Part I

Pasts

Figure I.1 Chickens explore. Copyright Catherine Oliver, 2018.

1 Relational animals

Beyond-human geographies require a shift in our imaginings, constructions, and lives. Rooted in relational thinking, this chapter troubles the deployment of binary thinking in both mainstream thinking about animals and in animal activism itself. In binary thinking, animal *beings* are constructed not only as sentient and feeling, but also hyper-individualised geographical subjects, to whom care is extended only contingently. Animal beings are most commonly found in companion animals and charismatic animals (Lorimer, 2007), and where animals have resisted their objectification, such as escaping from slaughter. On the other side of this binary are animal *things*; at best these animals teach us how to be human (a rhetoric also found in pet-keeping). The most visceral animal thing is the farmed animal: objectified, fragmented and, ultimately, consumed (Adams, 2010). This binary is historically contingent on violent interspecies hierarchies, and can also be found in animal activism based around rights or welfare. Veganism, however, attempts to make it possible to grapple with, and refuse, this binary to establish more equitable multispecies imaginaries of space.

Disturbing this binary requires a different kind of space between and beyond 'us.' Indeed, it requires a reconceptualisation of who counts as 'us,' to open new beyond-human relationships and spaces. This firstly requires a reconsideration of previous theorisations of animals, particularly those which primarily work through the expansion of anthropocentric rights-based frameworks. This chapter confronts these theories in, through and beyond the archive, a site itself of anthropocentric violence. In so doing, I seek to illustrate the permeation of binary thinking in animal activist histories and critique the persistence of the human/animal divide in animal activism, setting up the frame for understanding and relating to animals throughout the book.

Animal constructions

The condition of belonging, value or position in society is determined, for animals as for humans, across geographical contexts. For humans, race, citizenship, gender, disability, age, and more are determining factors in the degree to which particular people and groups of people are extended social, political, economic, and cultural worth, varying across space and time. Geographical scholarship offers a lens into these variances, and resistance to them. Beingness in this sense is not simply biological life, but is imbued with degrees of political, social, and ethical *mattering*. The extent of this mattering relies upon closeness to the white, Western subject who has continually, violently been constructed as 'human' (Jackson, 2020). The *propinquity*, here used to mean both the state of closeness and close kinship, to those constructed as less-than-human has had material consequences on the condition(al) of beingness, for humans and animals.

Pushing back against the violence of binary thinking in multispecies work is not new; troubling the human/animal binary is commonplace in critical animal studies scholarship. For example, attention has been paid to the categorisations attached to animals as 'pet, pest, profit' (Taylor and Signal, 2009), the privileges afforded to 'charismatic animals' (Lorimer, 2014), and the exceptional positions that companion animal species occupy (Sutton, 2020). The attachment of the condition of beingness or thingness to animals is historically and relationally negotiated (see Haraway, 1984; Butler, 1990; Braidotti, 1994; Irigaray, 1996). This chapter is not a call to shift the animal condition from animal things to animal beings, but rather to understand how the geographical and relational locations of animals determine their thingness or beingness, and instead ask how we might occupy instead states in-between being and thing.

The state of closeness of concern here is twofold: of being in proximity, as with companion animals; and of somatic markers of belonging to the privileged group of (most) humans (Cuomo and Gruen, 1998; Puwar, 2004). Interspecies violence is not historically unique in its mobilisation, but rather situates animals within 'interlocking and multiplicative systems of domination and submission' (Fiorenza [1992] in Pui-Lan, 2009, 193). The political and spatial deployment of human and animal is not oppositional, but a relational construct entangled with notions of closeness and distance, the familiar and unfamiliar.

Destabilising the being/thing binary risks reasserting problematic anthropocentric narratives by appealing to logics that reveal that if pain can be inflicted on those close, or similar, to the self, this pain could affect the self. Critiquing such an approach is not a matter of moral purity, but rather one of valuing difference, ensuring multispecies worlds because of their difference, not seeking to homogenise it in these appeals. It is only in moving away from modes of binary thinking between human and animal, and being and thing, and towards relational thinking based on contemporary geographical understandings of the body, relationality, and intersubjectivity that fuller understandings of multispecies worlds can be found.

In the binary mode, the character and intensity of feeling towards particular animals and groups of animals is an ordered and organised process, not an unhappy accident (Joy, 2009). The examples we commonly hear to contest human inconsistencies of loving and killing animals are at the global scale: cows, dogs, and cats are loved or eaten depending on where they are. However, the binary/thing plays out in more complex ways in the hyperlocal multispecies spaces. For example, in the companion animal relationship, a 'pet' is loved as an honorary human and welcomed, until they act in ways that disturb their expected servitude, such as biting (see Sutton, 2020).

Where a being/thing binary is often deployed in animal activism that sits farmed animals on one side, and companion animals on the other, this oversimplifies multispecies relationships. Geographical variation in loving and eating animal species reveals the construction of this value as socio-spatially produced between humans and animals. Importantly, however, what we must also attend to are the forms of (violent) closeness and claims of care (Giraud and Hollin, 2016) in killing, eating, and owning animals to unsettle and rupture this binary further.

The archive

The work in this and the following two chapters draws on archival research undertaken in the archives of Richard D. Ryder. Archival holdings and documents by their very nature as human collections of space, time, and events, force beyond-human realities out of reach. The stories of animals are rarely deemed more than marginalia to the affairs of humans to institutions that collect archives, and this is the case for the Ryder Papers. While it is an archive that features animals more commonly than most, it was primarily preserved to document the emergence of the politically and socially important animal activism movement in Britain: a movement of humans. The reality of the conditions of animals being fought against is easily subsumed by the affairs of the humans who are self-appointed advocates.

As an archival reader and listener, beyond-human research in the archives required reading along the archival grain (Stoler, 2009) in a space where animals fall under a regime of power which seeks to homogenise their stories as minor addendums to an all-too-human history. 'The archive lies not only in the clues it contains, but also in the sequences of different representations of reality. The archive always preserves an infinite number of relations to reality' (Farge, 2013, 30). Archival animals are as such 'temporally situated and yet also always in motion,' their existence 'constituted by stories of ... experiences, and encounters that might elicit alternative affective movements' (Lee, 2016, 34). The beyond-human stories in this chapter are entwined in politicised and expansive spatio-temporal networks of care; relations can shift with and in the archive, within an open but finite history (Nancy, 1990).

Before telling these stories, I will take a brief detour into how the archive is being constructed in this book. The archive is a holder of historical trajectories and contingencies: coming together and falling apart, interfolding. Understanding the interconnections and entanglements of there-and-then and here-and-now, archives 'always stand in active, dialogic, relation to the questions with the present puts to the past; and the present always puts its questions differently' (Hall, 2001, 92). The archive remains elusive, never complete, always reaching through, and allowing the present to arch back to the pasts that in turn weave into the future. These interweaved and relational histories are, even in the absence of animals, always multispecies; our lives coexist and reinforce one another and paying attention to beyond-human histories disturbs anthropocentric worldly narratives.

The language of discontinuity, of transformation, and rebuilding in relation to archival storage and reading (Foucault, 2002) resonates with an imagining of a break between violent pasts and possible futures imagined from a vegan perspective. And so, the past must also be confronted speculatively, in the mode of *as if* in order to centre the disruptive potential of relational history. In recognising the necessity of 'decentring human agencies, as well as remaining close to the predicaments and inheritances of situated human doings' (Puig de la Bellacasa, 2017, 12), we can begin to get into the beyond-human elements of history. How animals are included in the decentring of white Western humanity, however, must be approached with caution and sensitivity (Twine, 2014).

Anthropocentric theorisations and practices of animal rights foreground human superiority or beneficence, affirming humans' immovable centrality. This is evident in theories arguing for the inclusion of animals within a human framework, rather than a call to change the framework itself. Sentience and pain are the basis for the inclusion of animals in moral communities, most prominently through Singer's utilitarianism (Singer, 1975), centred on suffering as the moral impetus to do no harm, and Regan's deontology (Regan, 1983), which poses it as a moral right, or duty, to include animals in human rights-based frameworks. Ryder's painism combines Singer's aggregate suffering and Regan's conception of subjects-of-a-life (Ryder, 1983).

These theories do not remove the human framework but assume that animals can be moulded and anthropomorphised to fit human orderings. Pain and sentience remain within human conceptualisations of rights, duties, and human-centred worlds, ignoring non-sentient other-than-human ecological life and beyond-human community and interdependencies (Hall, 2011). They cannot offer transformative visions rooted in closeness and relationality.

Beyond binary thinking

There were several copies of the Speciesism Leaflet in the Ryder Papers (see Ryder, 2010) and each time my hands found it, or I return to look at it, the image contained shocks me. A tiny chimpanzee, curled over, weak, emaciated, and covered in scabs from injections of syphilis during experiments. I recognise pain, I empathise and his distress breaks through the archive, with a sudden kick of pain. The image, the first time I saw it, dragged me out of my almost trance-like state of searching archival materials (see Derrida, 1996) to the current moment, in a recognition of a shared feeling, of a desire to reach out to this body in pain. This chimpanzee exists beyond the archive and persists through the archive to reach into and affect my present. This beyond is a space that breaches history, geography, and species.

For Yancy (2005, 215–6), 'the body.is less of a thing/being than a shifting/changing historical meaning that is subject to cultural configuration/reconfiguration ... the body is a battlefield, one that is fought over again and again across particular historical movements and particular social spaces.' Through animal histories and biographies, some animals can (re)claim individuality and identity (Krebber and Roscher, 2019), but only this animal, not all. This animal becomes exemplary – the replica who reflects what we envisage of our own ideal image – in the rarity of their exceptionalism to 'the animal' (Derrida, 2005). This body breaches the binary. What is read in these images and stories is an animal who has – through historical and geographical circumstance – enacted their struggle towards the reader they did not know existed. We exist only in this relation.

It is within the *relation*, then, that negotiations and rejections of the binary are possible. This requires a critical unmaking of the self, a cultural reconfiguration of the body forging history, as not only geographically located, but geographically created (Yancy, 2005). An uncomfortable task, perhaps, but one necessary to critiquing beings and doings (Berlant, 2011) towards a new multispecies ethic. The role of relationality requires a recognition of our shared worldly embodiment (Weaver, 2013) at the same time as our difference. So, with the relation providing a space to undo the binary, what matters in the representation of this relation? How do we speak of our shared worldliness without overwriting our important differences? Representational power and voice are prescient concerns in multispecies work; when the powerful speak over the less powerful, they reassert structures of oppression and silencing (Spivak, 1988). Yet, it is a common rallying call to hear from animal activists that they must speak as a *voice for the voiceless*.

At first glance, a valiant effort to represent animals might appear in this statement. But the assertion of voicelessness is actually doing work to uphold animals as things, rather than part of a relational community that might speak beyond the language of the humans: 'consider that human language ... rather than separate from the animal, instinctively returns to it' (Massumi, 2014, 91). Complicity in upholding violence is an active process, manifested in a speciesist relation to and within the world. Articulating complicity *despite* intentions breaches our doing of relationships with animals, opening, and imagining alternative multispecies worlds. Turning back over a century, the question of voice and representation can be found with a brown dog.

In 1902, Lizzie Lind af Hageby and Leisa Schartau enrolled on a medical course at the London School of Medicine for Women. Here, they became one of the earliest undercover exposers of animal abuses in the name of science. In *The Shambles of Science* (Lind af Hageby and Schartau, 2012), the story of the brown dog is told. They first met the little brown terrier in December 1902, when William Bayliss undertook the first vivisection on the brown dog, cutting open his abdomen to demonstrate a medical procedure on his pancreatic duct. Two months later, Bayliss' assistant cut open the terrier again to inspect the previous wound, before clamping shut his abdomen and handing the brown dog over to Bayliss to cut open his throat and attach electrodes to the brown dog's salivary glands. Bayliss stimulated these nerves over the next hour as Lind af Hageby and Schartau witnessed the dog's body cut open, inadequately anaesthetised, clamped, and tied down to a wooden board, his mouth muzzled, in pain and struggling. The brown dog was handed to student (and future Nobel Laureate) Henry Dale, who removed his pancreas and killed the brown dog with a knife through the heart.

This is a story that tested Britain's perception of itself as a nation of animal lovers. The case went to trial. Three years later, a memorial was erected at a park in Battersea, inscribed: '*Men and women of England, how long shall these things be?*' infuriating medical students against anti-vivisectionist activists. This statue became the site, representationally and physically for questions of whose bodies matter. The brown dog was no longer killable, even though he had been killed; his body's memorialisation became the representative for political ideologies of anti-vivisectors. It projected a 'geography of suffering' (Garlick, 2015, 810) that violently disrupted modern conceptions of animals as beings or things, which depended upon the relation breaching enforced distance and exposing invisibilised practices (Arcari et al., 2020). Just as humans, animals and the world are entangled, breaching temporality and bringing human and animal closer through pain.

For Scarry, in pain, we are above pretence because of an 'absolute split between one's sense of one's own reality and the reality of other persons' (Scarry, 1985, 4) that cannot be reconciled until after pain or through a representative. This was the case for the brown dog (Figure 1.1). They became exceptional through their place in exposing a geography of suffering, after pain. Pain's world unmaking power persists in these collective spurrings to action, 'because the person in pain is ordinarily so bereft of the resources of speech, it is not surprising that the language for pain should sometimes be brought into being by those who are not themselves in pain, but speak on behalf of those who are' (ibid., 6). The brown dog was, at the time, the 'wrong kind of nonhuman' (Ingold, 2013). The brown dog's status as non-human afforded them a *lack* of care, an un-mattering. Pain overwhelmed the brown

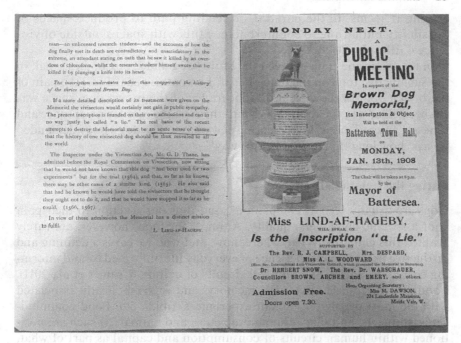

Figure 1.1 Brown dog memorial public meeting leaflet, 1908. Held at the British Library, The Ryder Papers [uncatalogued].

dog, in life as in death, but after and through their pain, a story is shared beyond impossible interspecies enunciations.

Spivak (1988, 33) asks 'are those who act and struggle mute, as opposed to those who act and speak?' The archival reader, located in archives as repositories of violence (Solis, 2018) has a responsibility to listen for those who cannot speak, or who are spoken for, to ensure that their struggle can be made to exist. It is not, in the case of the brown dog, that speaking was not attempted, but rather that the reciprocal of speech – to listen – is incomplete. This transhistorical listening occurs in between the then and now, where animal stories become agents in and of history, the present, and the future.

When we listen, we create conditions for the exceptional and close animal to speak, but animals who are not afforded space as narrators remain somatically and proximally distant. Such interventions rely upon the reconstitution of interspecies violence, where: 'constellations of violence that swell in one place are never constructed in isolation from other sites of violence' (Springer, 2011, 94). Pain's obvious somatic and representational power has always been deployed in animal activism, attempting to make a moral and political case to those inciting pain with varying successes and failures. The pervasiveness of animal pain and its image under these circumstances cannot, however, be without critique. We are caught, it seems, in the paradox of needing pain to produce spaces of possibility but failing to counterbalance this with spaces outside of violence (Gillespie, 2019).

In the space and time between the brown dog and I, our bodies speak in the subjunctive tone, in the tone of as if; I empathise as if I feel the pain of the brown dog. In the violent disappearance of the brown dog, space was changed. This dog spoke and was heard through the reordering of space and time as I entered the archive, creating new places. The dog's body demanded we ask 'how long things shall be this way'[1].

Eating the [relational] subject

Dogs are usually seen as companions, and thus our reaction to their pain is of a specific empathetic quality. Animals for food, however, are rarely granted this empathy. What, where, how, and who we eat is a defining and distinctive process and experience of everyday life, informed by and informing economic, social, and political circuits of culture (Cramer et al., 2011). Food is 'a receptacle of cultural memory' and is 'symbolically associated with the most deeply felt human experiences' (ibid., 59). Food communicates, connects, and shapes human societies, and animals as such are positioned within human circuits of consumption and capital as part of what, fuels humans. For food animals, 'their own bodies are created as the product of processes through production' (Wadiwel, 2018).

Thinking after Mol (2008), the theorisation of the eaten and eater's subjectivities always sustain one another. Eating is intimately entangled with embodying different kinds of culture and (animal) labour, but also ethics and politics (a thread which is picked up in Chapter 6). We must ask, then, whose body does my body consume to sustain me? The boundaries of the body are permeable, absorbing, and constituted beyond the contours of ourselves. Within these same bodies that we feel as so solidly defined, there are processes beyond our control. When we eat, our body is implicated with what (who) we eat, the processes of production of what we eat, and simultaneously is working beyond us, beyond our control. When we eat, our bodily and embodied choices, practices, and processes can be implicated in violence, or liberation, or many positions in between.

Food animals are viewed in the context of a necessary evil, rather than cruelty for cruelty's sake. As such they constitute an invisibilised geography (Arcari et al., 2020), kept out of sight (White, 2015), physically distant and psychologically divorced from the human consumer (Fitzgerald, 2010). For Arcari et al. (2020), drawing lines between the kinds of animals deemed useful for eating destroys wider attempts at multispecies ethics of care for 'nature.' Animal suffering cannot, however, always be forcibly distanced and breaching moments of somatic and physical closeness of suffering in unexpected and discomforting encounters (Garlick, 2015) in the passing of a slaughter truck; or when stray hairs, eyes, or faeces reveal the animal that was. I now turn to two historical campaigns that focus on the similarities between humans and animals. However, both fail to undo or resist binary thinking about animals.

In welfarist activism, the goal is to reduce animal suffering rather than ending animal farming. This kind of activism has long been palatable in Britain, meaning that even for more radical animal activists, an incremental approach has historically been prioritised, revealed in the great majority of archival materials in the Ryder Papers at the British Library. In welfarist approaches, the production of animals as things for consumption is upheld by its very opposite: through their beingness requiring they live and die 'well.' Welfarist campaigns appeal to animals' pain to ensure they have a good life, but do not contest that that life should be ended by killing.

One prescient example can be found in the Stop Live Exports campaign which was (and still is) a Royal Society for the Prevention of Cruelty to Animals (RSPCA) campaign focused on the transport of live animals between the United Kingdom and Europe, whose end goal was to slaughter food animals before transporting them to reduce their suffering. The transportation of animals was subject already to laws regarding the space afforded per animal, rest breaks, food, and water, but the RSPCA were suspicious this was not happening and as such trailed the drivers of the lorries transporting food animals (Saunders, 2002, 42–43). In their most recent statement on the Stop Live Exports campaign, the RSPCA call for the European Commission to cap transport of live animals at eight hours (RSPCA, 2017). The RSPCA claim because of the timings, this would effectively end live exports from the United Kingdom. In the 2017 campaign, they call on a 'nation of animal lovers' to end the suffering of animals on their way to slaughter.

In the case of live exports, the RSPCA are not engaging in disruptive or transformative practices, but rather exploiting animals' pain as temporally contingent to maintain their status as killable. Live exports campaigns reveal the navigations of the status of being and thing that seeks to trouble the temporalities and spatialities of animal death and suffering, but not necessarily to end it, nor to question its necessity altogether. Where food animals are lively commodities (Collard and Dempsey, 2013), their bodies and labour are at once being and thing, both theory and activism have appealed to shifting from one state to another.

The RSPCA position animals along this binary of being or thing, their beingness exploited in order to maintain and placate the desire to continue consuming their labour, bodies, and bodily labour. This upholding of binary thinking — human/animal, innocent/monstrous, consumer/consumed, being/thing — is perpetuated in examples of activism in the archives across all political iterations of activism. In particular, parallels can be drawn with the 'What's wrong with McDonald's?' campaign that led to the McLibel case, by now infamous in animal activist and environmentalist circles. The RSPCA are often attacked for 'partnerships with speciesist institutions, the explicit rejection of vegetarianism, and a desire to implement reforms that purposefully pose no threat to speciesism' (Wrenn, 2019, 90). Conversely, the McLibel case is viewed more generously by both activists and academics (Giraud, 2019) as enabling 'productive dialogue between elements of the campaign and theoretical work that has emphasized relationality' (ibid., 24) while not fully moving away from binary thinking of animals themselves.

Anti-McDonald's campaigning began with an action against McDonald's in the mid-1980s (see Wolfson, 1999; Morris, 1999). In the online archive, McSpotlight, the 'What's Wrong with McDonald's' leaflet can be found, which was produced and distributed by Veggies Nottingham[2]. In the leaflet, interconnecting issues of colonialism, capitalism, health and environmental issues are interweaved with the torture and murder of animals by McDonald's. McLibel presents the premise that liberation begins in your stomach, and that this liberation is not single-issue but rather connected with transnational, interspecies, anti-capitalist struggles.

> In the slaughterhouse, animals often struggle to escape. Cattle become frantic as they watch the animal before them in the killing-line being prodded, beaten, electrocuted and knifed. ... [McDonald's] sell millions of burgers every day in 35 countries throughout the world. This means the constant slaughter, day by day, of animals born and bred solely to be turned into McDonald's products. ... Some of them – especially chickens and pigs – spend their lives in the entirely artificial conditions of huge factory farms, with no access to air or sunshine and no freedom of movement. Their deaths are bloody and barbaric ... It's no exaggeration to say that when you bite into a Big Mac, you're helping the McDonald's empire to wreck this planet.
>
> (Veggies Nottingham, 1986).

The relational vision of the campaign highlights the entanglements of McDonald's violence across scales and space, and centre animal and food justice, from the individual human, economic imperialism, colonial invasion, to human rights violations of slaughterhouse workers (see Giraud, 2019). Despite the campaigns rooting in liberation beginning in the stomach, this relational vision doesn't extend to thinking animals themselves outside of the binary of being or thing. Animals are suppressed by this binary, their lives reduced to their bodies and their pain, forever cogs in a malevolent machine. While McLibel offers productive scaling out that situates veganism, importantly so, within wider liberation agendas, its relational power falls short of attending fully to animals within this.

Nevertheless, McLibel provides productive tensions for theorising the relational animal subject, in exemplifying the permeation of binary thinking also within radical grassroots campaigns. The manifestations of binary thinking about animals these different locations, ethics and politics reaffirms the need to return to and reconsider the non-innocent processes of caring for and living with, animals 'aware of troubling relations and seeking a significant otherness that transforms those involved in the relation and worlds we live in' (Puig de la Bellacasa, 2017, 83). Thinking thus about animals as relational subjects with whom differently navigating space and time can elucidate changing ethics and politics, how might we differently conceptualise this relationship, its fluid dynamics, and our collective navigation of the world and one another?

A politics of the possible

'If we take otherness to be the privileged vantage from which we defamiliarize our "nature", we risk making forays into the nonhuman a search for ever-stranger positions' (Kohn, in Kirksey and Helmreich, 2010, 563). Kohn's position of 'ever-stranger' might be interpreted across its two entangled definitions: as increasingly unusual, or as ever unknown and unknowable. If animals take the position of stranger, rethinking space from multispecies perspectives requires us to reconfigure the category of the stranger as we renegotiate distance and difference. The stranger exists outside of the opposition between friend or enemy, neither, yet embodying the possibility of both, and as such cannot be 'included within philosophical (binary) opposition, resisting and disordering it, without ever constituting a third term' (Derrida, in Bauman, 1990, 145).

If animals, human or not, are understood as *strangers*, they remain always neither/nor, exposing the fragility of binary thinking when faced with an undecidable or unknowable other; this is not an unknowable to be resolved, but the determinedly undecidable who threatens our epistemological grounds. Thus, figuring animals as strangers has implications not only for obliterating the opposition of being or thing, but the question of opposition at all and 'the very principle of opposition, the plausibility of dichotomy it suggests. [Strangers] unmask the brittle artificiality of division – they destroy the world ... They must be tabooed, suppressed, exiled physically or mentally – or the world may perish' (Bauman, 1990, 148–9).

Strangers become a problem when they refuse to stay strange, in the faraway, and instead demand to enter the worlds we acknowledge. Animals' constructions as strangers, especially food animals, are enforced and protected by the spatial distance of their lives and deaths from human spaces of living (Morin, 2018). It is only when strangers do not, by deliberate seeking out or chance, remain distanced that they threaten the foundations of the world by the potential exposure of processes of disembodiment and death. The threat to opposition, even when welcome, ruptures the social worlds that are defined by and for the collective self.

Where we find ourselves is always situated in the rupture; the relational self exists only in relation to this ambiguity. The stranger is embodied in and embodies the unknowable, threatening the foundations of binary worlds.

Reconstructing animals as beyond binaries, fluctuating socio-spatially, and historically contingent being/things requires an anti-hierarchical, anti-speciesist emancipatory agenda that not only demands the impossible (Perlman, 2018), but rethinks the boundaries of the possible to align with ethical and political beliefs beyond the human, towards animals as mutual constitutors in worldly transformations of the future.

History relies on an attempt to pause time and take a cross-section, as if by becoming completely still in the here-and-now, we can observe the past as static from our orientation, through an assumed linear temporality. A politics of possibility must be assumed in this multispecies, spatio-temporal work as part of the same political imagination that allows us to care *differently*. As Doreen Massey writes in *For Space*, 'not only history but space is open. In this open interactional space, there are always connections yet to be made ... relations which may or may not be accomplished ... space can never be that completed simultaneity in which all interconnections have been established, and in which everywhere is already linked with everywhere else ... This is a space of loose ends and missing links. For the future to be open, space must be open too' (Massey, 2005, 11–12). Because space and time must remain open – the former open-ended but finite while time is infinite, eternal, and immutable – the past can be read alongside a vision for the future. A 'relational politics for a relational space' (ibid., 61) entails interweaving with the beyond human, expanding our worlds not for, but with animals. Reading possibility backwards, through the archives, opens the beyond as a space for relational connections across times and spaces.

A politics of the possible allows us to live 'as if some experiences were evocative and reminiscent of others' (Braidotti, 1994, 5). In refusing animals as strangers, political priority must be given to reconsidering animals' representations as multiple selves, as between subject and object and embedded in networks of relationality of beyond-human worlds, through time as well as space. The beyond, then, is not only historical, but also the space of the future and for these new worlds, where we 'begin with the geography closest in – the body' (Rich, 2003 [1984], 212). An embodied approach to the multispecies allows us to rescale the relations between and beyond the present.

The multiple trajectories of animal and human lives are also stories of and interventions into space itself as an open, fluid, and transforming field on and in which to enact a politics of the possible. Imagining the future beyond the human requires opening 'a new figuration of subjectivity in a multidifferentiated non-hierarchical way' (Braidotti, 1994, 1). This way of thinking embraces in-betweenness, positioning the body as at once being and thing, never wholly subject or wholly object. The body thus is 'a point of overlap between the physical, the symbolic, and the sociological' (ibid., 4) through which we experience and enact living-together. The body is also where we shift between these states of beingness and thingness, fluctuating within this spectrum, both in and beyond the present, to the past, and to the future.

If we follow Bordo's conceptualisation of 'the body as not "me", but "with" me, [is] also the body that is at the same time inescapably "with me"'(Bordo, 2004, 2) then the world is understood as inhabited through a multiple 'withness' of the body. The body that is with me becomes a medium and a metaphor for cultural and social scripts; the body that is with me is the body that is read by the other, but it is not wholly me. This body is constructed through everyday reproductions, which become part of a double bind whereby being and thing (human and animal) are mutually exclusive, tearing the subject in two. Bodies consume bodies, as bodies are consumed – visually, socially, and physically.

This shifting between and beyond binary modes of thinking is the basis for rethinking animals throughout this book. I now leave behind these abstractions to delve into the archives further, asking questions about the tangible histories of animal activism in Britain, where questions of the body and embodiment are never far away. This chapter has set the scene for a reframing of animal life through the archives, entwining the past with the present, to which I return at the very end of the book.

Notes

1 This is the inscription on the original Brown Dog's memorial at Battersea's Latchmere Recreation Ground (1906–1910).
2 The anti-McDonald's campaign of the 1980s was organised by London Greenpeace and grew after McDonald's attempted to sue activists Helen Steel and Dave Morris who produced and distributed a longer anti-McDonald's fact sheet which led to the McLibel case (see Giraud, 2019, 23–26). This quote is taken from the longer factsheet (https://www.mcspotlight.org/case/pretrial/factsheet.html), not the shorter copy distributed on the streets (https://www.mcspotlight.org/campaigns/current/wwwmd-uk.pdf).

2 Genealogies of animal activism

In his presidential address to the Society of American Archivists, *Dear Mary Jane,* John Fleckner talked of how he found archival materials 'antiseptically foldered, boxed, and listed... wheeled out on carts, they were like cadavers to be dissected by first-year medical students... lifeless' (1991, 9). It was only once he became a 'would-be archivist,' working with archives before they were archives that their possibility (and his power) emerged: 'the records could speak to me in whatever voices my curious ears could hear, with whatever messages I could understand' (ibid., 9). This seductive archival mystery – what Arlette Farge (2013) calls the *'allure of the archive,'* Derrida (1996) experienced as *'archive fever,'* and Brent Hayes Edwards describes as 'something integral to the taste of the archive, to the sensation of encountering a past through it' (2012, 945) – is, for me, a befriending. This friendship in, of, and with the archive situates archival materials, figures, and stories within expansive and imperfect friendships beyond the archive.

Thinking of the archive beyond itself – as conceptualised in the previous chapter as a site of the *possible* – renegotiates space and time. In the archive, histories overlap, interfold, and affect the world. This and that are at once the two and the one. The singular and the plural. The here-and-now and the there-and-then are mutually dependent, in the past and in the future. 'The archival document is a tear in the fabric of time, an unplanned glimpse offered into an unexpected event' (Farge, 2013, 6), an ever-closer movement not just towards but beyond history. The archives 'tell of the truth' (ibid., 29), exposing the modes of people, relationships of power and producing meaning. Gabriel Solis asks of them, 'are archives of violence repositories of truth, or hope for the future, or both? Memory and truth have a contentious, if not volatile, relationship' (Solis, 2018, np).

Historical research centres a particular way of remembering, imbued with meaning by the powerful (Stoler, 2009), and archival memory is a human way of collecting, ordering, and storing politicised, classed, gendered, and raced pasts. The centring of raced and classed male subjects (Adams, 1994) has permeated animal activism across intellectual, organisational, and radical typologies, with white Western men being deemed legitimisers of feminised praxis of caring (Fraiman, 2012). This history of white, masculine rationality as a legitimiser of beyond-human care cannot be far from mind in the Ryder archives.

In this chapter, I contend that through their friendship, the so-called forefathers of animal rights have rewritten the histories of animal activism, and thus veganism. The work in the next two chapters is indebted to a long lineage of women and feminist animal thinkers including, but not limited to, Carol J. Adams, C. Lou Hamilton (2021), Brigid Brophy, Roslind Godlovitch, Amie Breeze Harper, Hilda Kean, Josephine Donovan, Lori Gruen, Ruth Harrison, Marti Kheel, Greta Gaard, Corey Lee Wrenn, Syl Ko, Aph Ko, pattrice jones, Sunaura Taylor, and many others.

Entering the archive: resisting powerful histories

Richard D. Ryder is a psychologist, ethicist, historian, and political campaigner, who grew up assuming animals were 'like him' and, as a child, seeing a dead blackbird on the street enlightened him to animals as sentient subjects, and the moral issues around ending or decreasing this pain (1998). Ultimately, he credits this encounter with beginning his turn towards animal philosophy and coining the term 'speciesism,' the concept for which he is most notorious, and 'painism,' his lesser-known (but more developed in his own work) approach to animal suffering. Ryder began using the term speciesism in 1970, claiming it came to him as a *Eureka!* moment as he bathed one day. Building on the work of British novelist and campaigner Brigid Brophy, speciesism attempted to define the animal cause as built on structural oppression and violence, comparative to other social justice movements, a project primarily developed in Peter Singer's *Animal Revolution*.

Ryder's archive offers a lens into the British animal activist movement of the 20th Century, but it is also inescapably Ryder's: a white, upper-class man's perspective and storage of history that reflects his privilege in even being institutionalised at the British Library. My time at the British Library invoked a particular sense of this history, as I was not working in the reading rooms, but within the archives themselves[1]. In the labyrinthine basements underneath the building itself, there is a seemingly endless capacity for curiosity and curiosities to be re-found, *brought to life* with the attention, hands, and words of researchers. Entering the labyrinthine basement where Ryder's archive was stored, the knowledge that this is just one basement of several compound my feeling of the infinite. How many lifetimes could I spend here? How many lifetimes are here? This wonderment, the impossibility, and the impenetrability of the Archives are bewildering. As I walked through rows of shelves, around corners, eyes catching glimpses of manuscripts and boxes, hands catching spines and dust, through to my own corner, *my* archive. Here, the archive came alive.

The 'assumption of a "politics of possibility" … is to imagine better versions of the worlds we inhabit, but also all the possible worlds that could exist' (Boyce Davies, 1999, 97). The possible worlds of the past and encounters with it are also the stories that define and transform the future and confronting these stories matters and shapes what we can (and cannot) imagine. Sylvia Wynter positions 'the human [as] *human narrans*' (2015, 25) – the storytelling species – and as such asks 'how are we not to think in terms of an ostensibly universal human history?' (ibid., 39). To story is to imagine and urge a world into existence and animals are often not far from our myths and stories (Stewart and Cole, 2014), affecting and affected by the retellings of the world deemed important enough to be immortalised. This thinking, representing or minding (Bekoff, 2002) is not just of humanness, but is also minding of earthly and animal. Entering the archive moved from a feeling of the infinite, to one of familiarity. The wonderment of the archives alienates, but this archive secrets me in, allowing me to befriend not only the archive, but the characters therein: 'the archive is an excess of meaning, where the reader experiences beauty, amazement and a certain affective tremor' (Farge, 2013, 31).

A metal box, faux-wood effect, pinned together with a plastic toothbrush, cased within a slightly larger cardboard box, calligraphy penned '*LJ Phillips Ltd. 139 New P—nt Street, London, 1*' and sealed with red wax; the box is sticky with dust. This box moved the archive from 'the haphazard, the contradictory, the potential' (de Leeuw, 2012) to the holder of a shifting and contingent history in the making. For Farge (2013), 'the reality of the archive lies not only in the clues it contains, but also in the sequences of different representations of reality' (30), in which this box can be momentarily (only momentarily) positioned as the holder of infinite imaginaries and possibilities of reality. Its potentiality and affective power, at the moment before opening, anticipates the absences, silences, and secrets that (may) emerge in an ethnography of and in the archives (Stoler, 2009).

The Ryder Papers were donated as an animal welfare archive, arriving to the library in 1999, and aimed to document 'the intellectual foundations and practical strategies of an important, and still growing, force within British pressure-group politics' (Summers, 1999)[2]. The archive has approximately 70 boxes of materials, consisting of correspondence, press cuttings, campaign materials from various organisations, memoranda, drafts of papers and books, audio and videotapes, and photographs. Part of the archive was un-embargoed in 2005, followed by a second tranche in 2015, but a third remains embargoed until 2025. Although the first two loads are now no longer embargoed, they were, at the time of research, uncatalogued and, aside from a cursory sorting, remained largely sorted as they had been filed and stored in Ryder's attic: messy, overflowing and filled with possibility.

Stuart Hall defined the archive as beginning at the moment at which: 'a relatively random collection of works, whose movement appears simply to be propelled from one creative moment to the next, is at the point of becoming something more ordered and considered: an object of reflection and debate,' and which 'represents the end of a certain kind of creative innocence, and the beginning of a new stage of self-consciousness, of self-reflexivity' (2001, 89). The first question asked of archives, what is an archive, must always precede (and follow) the question, what is *this* archive? At this moment, the whole apparatus of 'a history' as static and fixed becomes dissembled to construct the archive as a shifting, interconnected, and lively quasi-object (Latour, 1993).

In working with the archive, I respond in part to Ryder's own plea for animal activist histories to be formalised and institutionalised. Simultaneously, I question how far this formalisation is a reproduction of invisibilising and marginalising narratives of the *less powerful*, such as ecofeminist histories, in the pursuit of a linear narrative that excludes and obscures those without access to powerful institutions. In conversations with the archive and animal activists, I am pushed to ask, how can the archive speak back? How can I work in and from the marginalia to disrupt the anthropocentric archive? These questions were raised and struggled with in the previous chapter as questions of relations and constructions, but in this chapter, they are brought back down to earth.

Troubling vegan histories

In 1994, Carol J. Adams asked what, historically, has happened when a group who is supposed to be invisible tries to make animal issues visible (Adams, 1994)? Asking who is remembered, whose stories are told, and who is unnamed troubles spaces of memorial and constructions of history. Adams is concerned with the marginalisation and loss of women in histories of animal activism, and the potential plagiarism and erasure of their work, a concern echoed from this location of history-making. The rational and the masculine define mainstream animal theory, to the detriment of animals themselves and also destroying a genuine genealogy of the animal movement. By centring particular kinds of white male activists, there have been ongoing and repetitive attempts to legitimise the animal movement by distancing it from the embodied, the emotional, and the empathetic that is still prevalent today (Fraiman, 2012).

When thinking historically, what might have been is as important as what was (Stoler, 2009), and so is what was, but was not remembered. This is not a singular or linear history of veganism, but rather a foray into exploring and critiquing how humans have sought to ease or equalise the lives of animals in a world where it is more usual to dispossess animals of lives entirely. If veganism renegotiates the experience of somatic and proximal difference and distance between humans and animals, then it also has the potential to *break relations,* and remake them.

Veganism in Britain is situated within wider social and political life, and the idea of Britain as a nation of animal lovers. Contemporary veganism has its roots as a political and ethical movement dating back at least to the 19th Century (Kean, 1998). The definition of veganism most of us are familiar with today was proposed by the founders of The Vegan Society as 'a philosophy and way of living which seeks to exclude – as far as is possible and practicable – all forms of exploitation of, and cruelty to, animals for food, clothing or any other purpose.' How this definition came to stand as the official one is contentious; The Vegan Society's history says it was adapted from 1944 to 1948 – meaning there was no official definition for five years of the Society – before this definition was settled on. In 1948, Leslie J Cross was elected as a committee member, and he was determinedly concerned with veganism centring resistance to the [ab]use of animals. In The Vegan Society's Magazine, *The Vegan*, in 1949, he published an article saying that veganism: 'crystallised as a whole and not, as are all other such movements, as an abstraction. Where every other movement deals with a segment – and therefore deals directly with practices rather than with principles – veganism is itself a principle, from which certain practices logically flow' (Cross, 1949, 15).

Contemporary veganism is typically situated as staying as close as possible to this initial definition of veganism, combined with an anti-speciesist ethic most commonly attributed to Singer's *Animal Liberation* (1975) and Regan's *The Case for Animal Rights* (1983) that has been developed in homage and lineages following Ryder's initial definition. In 2010, Ryder argued that 'since Darwin, scientists have agreed that there is no "magical" essential difference between human and other animals, biologically speaking. Why then do we make an almost total distinction morally? The word "species," like the word "race," is not precisely definable. If we believe it is wrong to inflict suffering upon innocent human animals then it is only logical to extend our concern about elementary rights to the non-human animals as well' (Ryder, 2010, 1). Both the original definition of veganism and the more recent developments of speciesism are influential in contemporary veganism. Both are also defined within the rights-based, Western, masculinist authorisations of animal subjects through rationality and a rejection of sentimentality.

In studying history as continually remade and implicated in the present and futures (Nancy, 1990), the work in this chapter challenges also follows ecofeminist and veg[etari]an feminist traditions to resist this co-option of animal thinking and praxis (Probyn-Rapsey et al., 2019). Animal studies as a discipline is a relatively new and as such precarious discipline that has, over the last decade, become a 'legitimised' area of research in the humanities and social sciences, an increase in capital that has been linked to Derrida's 'authorization' of the subject (Fraiman, 2012). This authorisation is at the expense of feminist thinkers preceding the supposed founders and thinkers of animal subjects: 'why pause over a one-off talk by a thinker notably more concerned with words than flesh?' (ibid., 91). Peter Singer is widely attributed as the founder of the contemporary animal movement, after the publication of 'The Bible' of the animal movement, *Animal Liberation*, in 1975 (Villanueva, 2016), and is consistently cited by academics and animal advocates alike. Tom Regan is positioned as Singer's counterpart in bounding the Anglo-American-Australian field of animal studies between utilitarianism and deontological perspectives.

In 1976, Singer and Regan's edited collection, *Animal Rights and Human Obligations,* included one woman of seven authors, Marti Kheel, an ecofeminist scholar. Four years earlier, *Animals, men and morals* (Godlovitch et al., 1971) had been published, receiving high praise from Singer (1972), particularly Roslind Godlovitch's essay, *Animals and Morals*. This work of the 1960s and 1970s represented a turn in animal activism to ethical and political concerns (Singer, 1982). Animal philosophy between Brophy (1965) and Singer and Regan's collection (1976) was experiencing a growth and legitimisation, and it is not coincidental that this period also saw the disappearing of 'sentimentalised' feminist and women's work.

The animal movement as we know it today was founded on women's labour in the 19th Century (Beers, 2006), notably in Ireland and Britain through the work of women like Frances Power Cobbe, Louise Lind af Hageby, and Leisa Schartau. In the 20th Century, women such as Muriel the Lady Dowding, Brigid Brophy, and Roslind Godlovitch forged the field for contemporary animal thinkers. For these women, anti-speciesism was a social justice issue at the heart of political activism and action (Kean, 1998) and often was undertaken in conjunction with other feminist work for humans. These concurrent movements of justice, rights, and the personal as political influenced the animal movement even as society moved to increase animal suffering through industrialisation and large-scale animal agricultural practices (Harrison, 1964; Wrenn, 2019). Women's advocacy for animals was ethically and politically motivated by ideals of a more just society, but its belonging and practice by (wealthy white) women delegitimised its power, depicted as a past-time of those with little else to occupy themselves. This attitude persisted throughout the 20th Century, while at the same time the encroachment of male leadership was persistent.

As white middle- and upper-class men entered and claimed the animal movement, they legitimised their advocacy by deliberately distancing themselves from feminised concepts of love, friendship, and empathy, an attitude persisting into the recent history of the movement. This was expedited by the introduction of the concept of animal rights, which sought to expand the liberal human rights framework. In an interview with the British Library in 1998, Ryder himself expresses such attitudes that place in conflict the old animal activists, mainly women, and the new vanguard of young animal rights' men. Talking of those involved in animal advocacy in the 1970s, he described them as eccentric, old fashioned, humorous figures of old women in silly hats who were easy to laugh at and dismiss:

'I didn't know so many stupid people existed until I joined the RSPCA ...
Most of them were totally out of touch with reality, ineffectual people,
unpractical people. They were people who were the unemployed middle-
class ... they were – this sounds a bit sexist – the wives of people who were
earning enough money for them to have entirely free unemployed lives'
(Ryder, interviewed by Oxley, 1998).

Whether these women were incapable of campaigning or not is less
important here than the image of women animal advocates he describes
in a disavowal of sentimentality and feminisation in the animal movement.
Such a statement is symptomatic of a historical rejection of women from
the animal movement in order to professionalise (Wrenn, 2019) and legitimise (Fraiman, 2012) the credibility and authority of animal advocates to
work with animals in a feminised discipline by creating a patriarchal narrative around the subjects of the work, removing embodiment, emotion, and
feeling (Stallwood and McKibbin, 2019). This masculinisation of the field
affects women and other feminised subjects along race and sexuality lines,
whereby the work of indigenous, LGBTQ+, people of colour, and disabled
scholars and activists is regarded as somehow 'lesser.'

Not contained to academic work, this also affects and shapes veganism
itself where emotion is irrational and worthless, except in the development
of *affective activisms* that remain reliant on constructing the rights of
animals. It followed, therefore, that 'love doesn't come into it.' (Stallwood
and McKibbin, 2019, np), and emotional relationships with animals are
rejected as a serious mode and motivator of transformation, embedding a
rejection of feminine subjects in the movement. However, in recent history
there has been some evidence of the return of these feminised sentiments
in animal academia and advocacy being practiced by some white male
thinkers and advocates. The decades long distancing of feeling now being
reclaimed and legitimised by white, Western male subjects is a worrying,
if ironic, concern (see Wright, 2020) that might further push out marginalised humans.

Class politics have also haunted the animal advocacy movement, where
critique of the representation of the animal movement and veganism has
long been as a wealthy middle-class lifestyle (Morris and Oliver, 2016).
Veganism holds close associations to whiteness, embodied in the figures who
are constructed and upheld as its leaders. This is not necessarily reflected in
the constituency, with African-Americans being the fastest growing demo-
graphic of vegans in the West (McCarthy and Dekoster, 2020). The history
of animal studies and animal advocacy are not separate but entwined, with
gender divisions in the constituency of vegans and animal advocates similar to the gender split in animal studies academics[3] (67% female, 30% male,
3% other for academics, Probyn-Rapsey, et al., 2019). The actual and perceived whiteness of the animal movement is related to attempts to create a
'race-neutral' movement (Harper, 2013) that is critiqued by black feminists
who ask how being racialised and sexualised within the animal movement

creates a different type of vegan praxis (Harper, 2010). The social, political, and cultural dynamics of power and identity are not eschewed in vegan thinking or practice, but rather are reproduced in the dominant narratives of historical and contemporary stories.

The history of the animal movement

The 'legitimisers' of animal activism – Derrida (2002), Singer (1975), Regan (1983), and Ryder (2010) – have undoubtedly shaped the kinds of discourse and action being deployed in contemporary veganism. However, most animal work and animal thinking has been (and continues to be) undertaken by women (Probyn-Rapsey et al., 2019). This is thus a history of silencing of the events, people, and organisations whose marginality is not even on the marginalia. These feminised subjects include animals themselves, whose 'absent presences' (Adams, 2010) in activist work are evident where they are referred to, not brought in. Memorials of animals in the archives are similarly hidden. Reading along the archival grain (Stoler, 2009), historical movements inform understanding and critique the practice and networks of veganism and of human-animal relationships through *friendship* as a practice of exclusion, solidarity-building, and historical narrative claims.

Brophy's *The Rights of Animals* was the beginning of the transformation of the animal activism movement of the first half of the 20th Century, informing the emergence of ethico-politically engaged anti-speciesist contemporary veganism. The lineage of animal activism has been covered elsewhere (Ritvo, 1984; Kean, 1998; Steiner, 2005; Phelps, 2007; Preece, 2009). Rather than reproduce these histories, I seek to respond to and disrupt these dominant histories through the archive to rethink whose history is represented, whose activism is centred, and how read this through *friendship* as an inclusionary and exclusionary bonding in three kinds of activism: academic, organisational, and radical[4].

An academic, conceptual, or theoretical concern with animals as co-creators and partners in the world may be traced in Britain as far back as Bentham (1780 [1982]). Brophy (1965) at once criticises and offers human-animal relationships as a thought experiment in morality and urges abstractions of theory to be coupled with praxis (as quoted in the introduction of this book). The influence of the work pioneered and influenced by Brophy remains central to many contemporary animal thinkers today. Academic activism is not only those within academic institutions, but more broadly encompasses thinkers, authors, and practitioners who further thinking with and for animals as mutual constitutors of the world and with a directive of living with animals and the world less violently. Rather than policing the academic as that within university or educational institutions, academic activism is far broader, expanding this constituency to those who work outside of traditional institutions.

Organisational activism bounds a wide variety of organisations, which may be small or large, regional or global, and are here understood as an aboveground, co-ordinated, and accountable group working for animal activism. Organisational activism may follow gradual change or call for immediate abolition. Within this category, a full spectrum of approaches, beliefs and structures may be visible, and at either end of the spectrum may overlap significantly with academic activism. These organisational activism groups are typically public facing and often accountable charitable organisations. Thus, they do not engage in legally dubious activities, although they may offer support to these activists.

The final constituency is radical activism, which historically has a consistent and important role in direct action, lobbying, and urging forwards progress for animals, whether through protectionary or liberatory tactics. Crucially this is emerging with vigour in contemporary veganism, enlivening a veganism that has been perceived of as more sedate in the mainstream post-1990s revelation of undercover spy cops within activist cells and groups, both vegan and wider social justice groups, arguably the most effective legal transformations for animal welfare (Bourke, 2019). The movement of this undercover work from private to public breaches the attempts at secrecy of the violence of eating animals and its use in legal cases (usually through collaboration with organisational activism) in turn legitimises their actions in public discourse.

Over time, these categories have become ever more intimately associated and dependent on one another. Reading histories across these categories provokes, explores, and understands animal activist space as messy, entangled, and constructed with and by social relations, particularly of friendship. In order to traverse these different activisms, Ryder's archive is in conversation with other archival materials.

Animal activism in the archives

The Oxford Group were an academic collective, of sorts, formed of Oxford postgraduate students largely of philosophy and sociology, founded by Stanley Godlovitch, Roslind Godlovitch, John Harris, David Wood, and Michael Peters (see Garner and Okuleye, 2020). Their formation was sparked by Brophy's *The Rights of Animals*. Ryder wrote three letters to the Daily Telegraph in response to Brophy, who then introduced him to the Godlovitch's and Harris, as they were all living in Oxford in the late 1960s. Within a short time, they were distributing anti-vivisection and anti-hunting leaflets and organising small protests, inviting activists from London to Oxford to disrupt an otter hunt (1971). The rebranding of animal activism away from older women and centring young, white men of some social standing, Ryder (1998) argues, was essential to its progress and being taken seriously.

In 1970, Ryder 'coined' the term speciesism and circulated the 'Speciesism' chimpanzee leaflet (discussed in the previous chapter, and reproduced in Ryder's *Speciesism Again,* 2010) around Oxford, which connected species with race and gender, claiming this as his original idea. The Oxford Group was at this time joined by Peter Singer, with whom the idea of speciesism quickly became a central concept to develop. Speciesism has been cemented within academic animal activism and specifically vegan studies (Phelps, 2007) over the fifty years since, and has recently become popular in activist circles. Thinkers of the Oxford Group, notably Peter Singer, claim this enabled ecofeminist and anti-racist thinkers to further their theories. However, Anglo-American feminist thinkers had been writing about animals for at least a decade before his publication. Adams' first essay on feminism and vegetarianism was published at the same time (Adams, 1975) and she was already establishing more complex theories and exposes of the intersections of race, gender, and eating animals that cannot be accounted for in Singer's utilitarian approach.

Animals, Men and Morals (Godlovitch et al.) was published by members of the Oxford Group in 1971. It set forth a mainstream animal agenda that persisted for (at least) the next two decades, centring the voices of powerful white Western thinkers within or aligned with the Oxford Group: 'Once the full force of moral assessment has been made explicit there can be no rational excuse left for killing animals, be they killed for food, science, or sheer personal indulgence. We have not assembled this book to provide the reader with yet another manual on how to make brutalities less brutal. Compromise, in the traditional sense of the term, is simple unthinking weakness when one considers the actual reasons for our crude relationships with other animals. to argue that a lack of compromise is wrong-headed is merely to perpetuate various fantasies people have about the regard that should be had toward other species' (Godlovitch et al., 1971).

Their work in this book was largely ignored at the time, leading Singer to review the volume (1972), where he first put forth his utilitarian arguments for animal rights. This review has been positioned as the predecessor to Singer's 1975 *Animal Liberation* (Phelps, 2007), further cementing the narrative of the birth of a new kind of social movement for animals. Singer (1982) describes the Oxford Group as semi-communal, with philosophical disagreements but a mutual interest and agreement on the 'unacceptable' treatment of animals. He describes Ryder as his mirror image: where Singer was a vegetarian who did not oppose medical testing, Ryder was staunchly anti-vivisection but not a vegetarian. The two's friendship has a powerful hold on the history of animal activism told and remembered in both activist and academic circles.

Ryder and Singer were aware that histories of animal activism preceded their own but claimed that their legitimising voices moved animal activism from the sentimental concern of ladies to a political – and rational – cause (Ryder, 1998). In Ryder's archive, a handwritten letter to Ryder from

Christine Stevens, President of the Animal Welfare Institute (USA, 1998) writes to Ryder: 'I note your sentence "we simply do not know what went on in the US in the 1940s, '50s and '60s. AWI and SAPL were very active in the '50s and '60s–well before the book *Animal Liberation* was published. SAPL and AWI always had a low profile, so I think your impression that "everything took off in the U.S. and that before then it had a low profile" doesn't reflect the actual situation.'

Stevens goes on to suggest that *Animal Liberation* inspired grassroots collectives over the country, but that large organisations predated the attention of the Oxford Group's members, with much longer histories. While claims over the transformation of animal activism in the '60s by Ryder, Singer, and their friends might well have been valid as a change in direction, their narrative sought instead to erase their predecessors. Nonetheless, it is the Oxford Group's story that has successfully dominated Anglo-American animal activism. The legitimising actions of the friendship of the Oxford Group as elite activists is at the expense of erasing activist histories. This suggests that this kind of *friendship* that destroys all it excludes, extracting others' knowledge to claim as their own.

At a similar time, in the early 1970s, feminists were adopting vegetarianism through non-violent, anti-war activism. As ecofeminism began to emerge in conversations, women connected with the earth and the earth and animals' exploitations were understood as intimately entangled with women's own societal, political, and cultural struggles against masculine consciousness (Ruether, 1975). Feminist thinkers, like Brophy (1965) and Harrison (1964) had published these ideas at least a decade before Singer and Ryder entered the scene.

In *Ecofeminism,* Adams and Gruen (2014) explore how 'feminists were expressing formative theoretical and practical insights' (ibid., 2) since at least Edith Ward in the late 19th Century, through the 1970s non-violence activism and into the 1980s with the emergence of a particular kind of feminised care for animals through sanctuary (Davis, 1989). These intersections were elaborated long before Singer's (1975) white masculine utilitarian ethics, which contradict the practical ethics of feminists working with other animals. Yet, this continued centring of animal *rights* theories – specifically Singer's – as the cornerstone of animal thinking has legitimised their hold on a school of thought, in both the past and present. In the Ryder Papers, these feminist histories were all but invisible, found only subsumed in the narratives of great white men.

While in 1970s Oxford, academics were organising, thinking, and writing about animals, human duties to them and their rights, in 1970s USA, feminists were elaborating and practicing resistance to complex oppressions across species through non-violence and un-learning masculinised consciousness' of the earth and other animals. Simultaneously, organisational animal activism, traditionally welfarist, was also facing challenges from the radical margins.

In the longest standing animal welfare group in Britain, the Royal Society for the Prevention of Cruelty to Animals, there was an activist intervention into the conservative Society, by the Reform Group. This group was concerned with ousting pro-hunting members of the RSPCA and addressing the organisation's failure to deal with factory farming or animal testing. The Reform Group promoted and nominated members of their group to the council to represent more radical positions in the conservative organisation. The struggle between the reformists and the RSPCA continued with a similar group, the RSPCA Action Group (RAG) forming in 1985 when the conflict of the Reform Group had largely died down. The radicals of the 1970s had mostly left the RSPCA (Ryder, 1989) following the 1974 *Sparrow Inquiry* into financial mismanagement and RAG members followed in footsteps of the earlier Reform Group trying to enact radical change in the RSPCA from the inside (Garner, 2004).

Radical activism was also undergoing a transformation in 1970s England, towards direct action and liberation tactics, most notably under the Animal Liberation Front (ALF). The rules and membership of ALF are not a definitive group, but rather cells of activists working clandestinely either alone or in small groups. Band of Mercy[5] was one of the more notorious cells, created in 1972 to organise these actions. It began as a hunt saboteur's group, but quickly broadened to target other animal exploitation sites, particularly seeking out economic targets. Band of Mercy claimed responsibility for the first known act of arson by the animal liberation movement, in November 1973 at Hoechst Pharmaceuticals (Milton Keynes). Animal liberationists have been assaulted and jailed for their legal tactics of hunt disruption and this has led some activists to seek more effective, albeit illegal, direct actions.

One of the Band of Mercy organisers later left the liberation movement, while the other banded a new militant group – the aforementioned Animal Liberation Front – in 1976 (Best and Nocella, 2004). The latter served two prison sentences for his direct action, for property damage (1975) and liberating mice (1977). The rise of these kinds of tactics in the mid- to late-1970s led to heavier surveillance and policing and as these tactics spread, the ALF had become an international umbrella for animal liberationists. These images of animal activists continue to permeate mainstream images of activists, such as the use of actual ALF images used in the Netflix film *Okja* (Joon-Ho, 2017) and these radical histories informing the film's narrative (Parkinson, 2018). As a decentralised group, with no organisation or membership, there is an interest in how we can trace these underground histories (see Roscher, 2009), where activists do not know one another nor necessarily any simultaneous actions being undertaken. The risks of maintaining archives pertaining to actions is a risk of incarceration or punishment as far as it is one of exposure of continuing activism.

The temporal simultaneity and interaction between the academic, organisational, and radical are introduced in the (very) brief history of activism from the archive shared here, connecting differently oriented

and geographically located activisms in this era of emerging transformative animal activism, in which contemporary veganism is rooted. These
entangled lineages of activism remain present under the umbrella of animal activism and within veganism today. Pressingly, particular forms of
radical activism remain idealised in contemporary veganism in upholding
the 'ideal activist' as untied from society, free of responsibility, unwilling
to compromise but willing to sacrifice (Craddock, 2019), ideas which have
echoes in contemporary veganism, discussed in Chapters 4 and 5.

Women in the margins: Roslind Godlovitch and Brigid Brophy

In the Ryder archive, women's work was found only through their intersection, association, and friendship with 'archivable' men. The 'pussy panic'
of and within animal studies defines its histories and praxis, informing the
legitimisation of the field by aligning to masculinist rationalised work, centring Derrida and 'the male scholars who follow him, using his work to
build an alternative genealogy for the field' (Probyn-Rapsey et al. 2019, 199).
This masculinist turn is perhaps, then, not so much a turn as inherent to the
field from its inception, seeking institutional, and disciplinary acceptance
by distancing from feminised work, creating both ideological and structural
pressures on women and women's work. When a group who is supposed to
be invisible tries to make an issue visible, the association of the issue with
the invisible group diminishes its value.

Feminist readings of the archive are necessary to working with and sub-
verting the narratives of the powerful as the only possible histories. Ryder
was and is welcomed by institutions; his knowledge of archiving, collecting,
and protecting his own history as well as that of the animal movement is
not divorced from his elite position. Yet within his archive, there are others circulating within and around his narrative, particularly through his
many friendships that reveal starkly the problems of invisibilisation and the
necessity of networks of friends in 'leading' the animal movement. None
was so formative perhaps as the friendship Ryder had with Brigid Brophy,
who brought him into the public-facing animal movement and connected
him with other his group of activists in the 1960s.

As I wrote in my field diary, 'Brigid Brophy has a large presence in
Ryder's archive, in correspondence cited as an inspirational figure to
Ryder. In the archive, I found a copy of Brophy's *The Rights of Animals*
which, according to Ryder's correspondence, was the inspiration for his
own activist tactics in the 1970s, particularly those opposing bloodsports
within the RSPCA. But who is Brigid Brophy? Have I become so familiar
with her name and her presence in Ryder's archive that I have imagined a
whole history for her? A pamphlet for *The Brigid Brophy Conference 2015:
A Felicitious Day for Fish*. She's everywhere but nowhere' (November 2016).

Brophy was spoken of fondly by Ryder and their correspondence was
extensive. She was one of few women present in the archive. Contemporary

veganism's representation of its white male middle-class leaders is a clear historical trajectory, despite veganism's feminisation and the high proportion of women representationally in veganism (Probyn-Rapsey et al., 2019). Disrupting the history of the powerful by searching for the still powerful but side-lined women proved only a partially fruitful approach to seeking a vegan history that more accurately represents its constituency. Brophy was also a powerful figure as a white woman of great social and economic capital; it is unsurprising to find her in the British Library, and she no doubt features in several collections beyond the Ryder Papers. Hierarchies are not being destroyed, but perhaps slightly disrupted.

Brophy's *The Rights of Animals* (1965) has been taken as the moment of creative construction of the ethico-political animal activism movement, notably by animal activist Kim Stallwood (2013). Her intellectual legacy informs contemporary veganism from this moment, but Brophy's contributions to ethical and political animal thinking are largely unrecognised as deliberate disremembering and disinheritance of the feminist history of animal activism. In 2018, English Heritage deemed it 'too soon' to recognise Brophy's contribution to history[6]. The power of archiving, memorialising, and remembrance attaches significance to historical contributions being entangled and presupposing awareness of that contribution *through* memorialisation (Grever, 1997). Centring Brophy is not only to remember women's presence and establishment of animal work, but to centre her contribution to defining the contemporary animal movement as the beginning of a movement away from what was a fractured disorganisation towards a coherent collective.

> In all relationships between humans as a group and the other animals as a group, the moral position of the humans is straightforward. We simply override the others. Arbitrarily and wantonly-without, that is to say, justification or necessity-we arc tyrants. In effect, we make the other animals into a slave class. Our behaviour simply ignores their rights.
>
> *(Brophy, 1971, 125–126)*

Brophy's essay was adapted for publication in *Animals, men and morals* (1971). In this book, traces of other women written out of the intellectual animal tradition can be found, most notably Roslind Godlovitch. Godlovitch was central to the Oxford Group, yet is little known. Godlovitch was an original member of this Oxford Group, alongside her partner Stan and John Harris (see Garner and Okuleye, 2020). She provoked a certain curiosity and elusiveness in the archive, despite her potentially transformative role in the history of animal philosophy. Peter Singer suggested that it was Godlovitch's influence that convinced him of the vegetarian position, acknowledging that 'so many of [his] ideas had come from others, and especially from Ros, [he] should allow her to publish them' (Singer, 1982, 8). Singer's impression of Godlovitch was that she was intellectually admirable, politically astute and held a firm belief that her work, along with the others of the Oxford Group, may trigger a widespread protest movement. Yet the masculinist politics and events of the emergent ethico-political animal movement, meant that Singer's work remains so incredibly influential, while Godlovitch has disappeared (almost) without a trace.

Godlovitch's 1971 essay *Animals and Morals* is formative in proposing the possibility of imagining part of the way the lives of animals, seeking not transcendence but a shared embodiment and being in the world that follows imagination: I think the confusion in this contention [that one cannot imagine being an animal] lies partly in overestimating the amount of oneself that needs to be taken along in the act of imagination and partly in an ignorance, or blindness, which most of us share about animals (Godlovitch, 1971, 23). Finding the traces of women mainly in the margins of perhaps Britain's largest and certainly most prominent animal activism collections is symptomatic of a movement that remains controversially entangled with issues of race, class, and gender. Animal activism purports its moral and political grounding on principles of anti-racist and feminist that it has yet failed to fully grapple with in its own centring of particular (white, male, Western) voices.

Beyond a singular history

Animal activism's history is never singular but rather must be read across radical, organisational, and academic streams, through the movement of people, ideas, and actions. What lies beyond – the archive, my knowledge, this location – offers a rich potentiality of partial yet still multiple pasts of activism for animals. For Ryder (1998), his own relation to these (his) pasts is one of *nearly* getting there. He believes the tide is always going against animals and persistence is the key to achievement, along with stamina for disappointment and attacks from one's own side. Whether Ryder would apply these attacks to the exclusions and rewriting of histories from his own group is uncertain. In this chapter, I have begun to tease out the entanglements and allegiances within animal activism in the 20th Century through insights from the Ryder Papers. This chapter has critiqued the oversaturation of masculinised animal activist histories. In the following chapter, I trouble these narratives to theorise a friendship within and beyond the archive, posing this as resistance to dominant narratives of these histories.

Notes

1 I was working in the British Library as part of their 2016/17 cohort of PhD placement scholars, under the supervision of Dr. Polly Russell and Gill Ridgeley. To them, and the BL itself, I am deeply thankful for the opportunity, space, and trust to work in this unique and intellectually stimulating way.

2 Letter from Anne Summers (Curator in Dept of Manuscripts) to Ann Payne (Head of Dept Manuscripts); 5/2/1999.

3 Disciplinary differences may also be of note: philosophers were 61% male, and politics 44%, disciplines particularly implicated in reproducing hegemonic masculinist work and renowned for damaging environments for women scholars.

4 Corey Lee Wrenn's *Piecemeal Protest* (2019) offers a fuller account of animal activist and vegan organisations, factionalism, and progress of non-profits.

5 The name *Band of Mercy* was revived from a 19th-century RSPCA youth group.

6 See, for context, a 2018 tweet from English Heritage arguing it is 'too soon' to recognise her historical contributions, but that these may become important enough to honour in the future. https://twitter.com/EnglishHeritage/status/1057268436936679424

3 Beyond-human geographies of friendship

For Georg Simmel, negotiations of distance and difference are counterbalanced by a desire towards *commonness,* whereby relationships or friendships are not exceptional but rather necessarily expansive in their absolute everydayness. Quoting from *L'Etranger*:

> in their relation, after all, they carry out only a generally human destiny; that they experience an experience that has occurred a thousand times before; that ... they would have found the same significance in another person. Something of this feeling is probably present in any relation, because what is common to two is never common to them alone but is subsumed under a general idea which includes many possibilities of commonness. No matter how little these possibilities become real and how often we forget them, they thrust themselves between us like shadows.
>
> (Simmel, 2008, 326–327)

This chapter focuses on encounters of and in the archives, tracing particular stories of animal activism to construct a trajectory of friendship beyond the archive, positioning the enduring pulse of friendship as shaping contemporary networks of veganism in Britain. Rooted between an ethic of care (Gilligan, 1982) and a politics of the possible (Braidotti, 1994), archival stories focus on how through our relations and in relation, worlds have been constituted. For human-animal relations, this scales up and up, down and down (Lorimer, 2003), in dynamic correlation with the world, calling for a politics of the small, the personal, the relational, the everyday (Stewart, 2013). The everyday, more so than exceptional events or extraordinary moments, has shaped the vegan movement between human actors and by extending friendship to non-human animals.

Derrida (2005, 1) critiqued the 'major' (white, male) figures of Western philosophy's treatment of friendship as always emergent, always *to come,* and never realised: 'O my friends, there is no friend.' However, a friendship outside of white masculine Western imperialist philosophy is never allowed in this 'one true friendship' (Montaigne, 2004 [1580]). Friendships that uphold elite and powerful have dominated animal activist histories. However, with a feminist ethic and praxis, friendship is instead understood outside of the primary (usually heteronormative, cisgender, monogamous) romantic relationship and (nuclear) familial structures. Friendship in the archives excludes and includes, intentionally, within networks that remain open and expanding. The realm of friendship is typically positioned as supplementary or replacing primary romantic or familial structures, but rarely as central itself (Kern, 2019). A revolutionary geography of friendship *as* resistance has not yet been mapped as a socio-spatial relationality that refuses to be bound or shaped by hierarchical power (Levine, 2012). Where powerful individual, group, and even state relationships are named friendships, this works to protect their position in social and political hierarchies. To refute these as friendships means rejecting Western imperialist friendship as always *'to come'* and instead centring something somehow different.

Where 'friend' has its etymological root in the Germanic base of 'free' in the sense 'to love,' 'friendship,' the state of being a friend, has its base in 'freedom,' the state of being free (OED). Returning to this root of friendship is to refuse its use by the powerful to protect their position and instead return to those seeking and working for freedom beyond themselves. In this chapter, I move towards friendship as a feminist ethico-political approach and practice of freedom in the relations between humans and humans, humans and animals, animals and animals. It is through these relations which intra- and interspecies bonds might be constituted, sustaining themselves, and resisting their own co-option.

Befriending the archives

Historical friendships reconstitute understandings of vegan histories, as well as elucidating how these relationships and approaches of, to, and with friendship are necessary to understanding ethico-political veganism between humans and extending this to animals. Searching for archival animals whose presence has been absent in the human archive, this chapter builds out of a loss and disconnection of fleshy embodied realities, in the endless objects of the archive. This overwhelming weight of the past, and the storage of the past, means this chapter pulses towards an end of commonness – what is common to the two is never common to them alone – not only as a generally human destiny, but a destiny of possibility between and beyond an expanded *us*.

In March 2017, in my archival field diary, I wrote about a letter: 'As I write, a letter waits for me or, I suppose, I wait for me to go to it. This letter, both before and after I open it, is a holder for the friendship to, of and beyond the archive. The letter breaches a boundary, between what is and what could be. When I open that letter, the event of opening and reading that letter, shifts the networks and relationships I have and have imagined.

That letter moves me from reader of the archive, to being archived. It is not only concerned with friendship and silence, but it transforms the very question I am asking, that I am being asked by throwing into question the spaces and histories that I am implicated in. The letter is full of meaning, both tangible and intangible; said and unsaid. This is a story of the archive, beyond the archive, of being archived, where the archival bond reaches me, and I become part of this history that continues.'

The letter was a reply to a short article I had written about Ryder, his archive and its importance to animal activism histories (Oliver, 2018) which I had sent to him, along with asking for permission to reproduce some of the materials in his archive for my research. I had been working in Ryder's archive for six months and I felt as if I had come to know him. In his archive, I had watched him grow older, suffer, succeed, along a journey of almosts within and beyond his activism. But most of this knowledge is mine alone, engaged already in a careful curation and selection processes: what did I photograph, analyse, and share? The researcher's agenda and orientation are not neutral. Ryder's pre-catalogued archive had, at this point, been seen in depth only by me, and this letter is the container of potentiality and possibility, imbued with meaning. Its contents offered the return of friendship or the destruction of my befriending.

If we 'must start with the friend-who-loves not with the friend-who-is-loved' (Derrida, 2005, 9), this means beginning with the self who is aware of the active loving, or the active/act of friendship, rather than whom is loved. Because friendship emerges from the lover, this 'loving belongs only to a being gifted with life or with breath ... being loved, on the other hand, always remains possible on the side of the inanimate' (ibid., 11). The archive may be inanimate, kept alive, and lively only by the befriender, but the feelings and emotions invoked, and the relationships formed with archival animals and humans then precede and follow the knowing of these characters as existent and real beyond the archive. A friendship drawing on feminist relationships to reshape, reform, and resist linear and dominant friendships and histories might hold accountable reassertions of oppression across human and animal collectives. Friendship might map an open and elongated spatial and temporal network of people connected by caring for animals and as an analytical tool to resist the already recurring co-option of veganism and caring for animals by rational, masculinised structures.

The space of friendship

In the previous chapter, I shared the story of the brown dog (Figure 3.1), whose pain breached the archive as his body was cut open. I want to briefly return to where we left the dog, memorialised, as the site representationally and physically over the question of 'whose bodies matter?'. Riots,

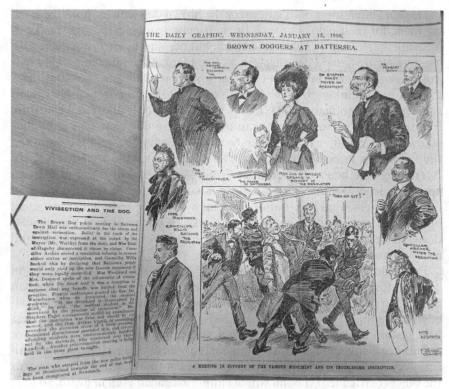

Figure 3.1 Drawings of 'Brown Doggers at Battersea' trial in The Daily Graphic, 1908. Held at The British Library, The Ryder Papers

vandalism, and the need for 24-hour police protection led to, in 1910, 120 police officers removed the statue from the park in the middle of the night, supposedly being melted down to nothing.

This archival encounter speaks of the way bodies do, or do not, matter in questioning who, how, when, and where we care for. Connecting through and beyond the archive, the Brown Dog affected and stayed with me. I held them. And I am not the only one (see Kean, 2003; Goldman, 2010; Garlick, 2015). The brown dog's body story, and the women who took a risk to expose the cruelties of medical demonstrations on animals, stay with me. The dog's body pain speaks after the moments of pain itself (Scarry, 1985), through re-tellings of their experience and eventual death, their memori-alisation, and the resultant disruption of public space not only asks whose bodies matter, but how and to whom. The brown dog cannot have known their pain would speak in these echoes; they would not have felt the care of hundreds opposing this treatment as they waited, caged, their next incision. They would not have known their body breached space and time for me, for us, to hold them repeatedly.

115 years later, some of us still carry the brown dog with us, holding space for them, for their pain. Through the befriended archive, the dog speaks to me, with me, for me. I cannot save them, but perhaps by carrying them with me, by sharing them, I can save others like them, others like me. This holding space is filled with ethico-political imaginaries and possibilities, necessary for (re)thinking friendship with different (human-animal-historical) subjects. For Montaigne (2004 [1580]), there can be only one true friendship, leaving all other relations as acquaintances, or familiarities. This unboundedness of friendship can be understood as extending beyond the events of the friendship in both directions, prior to and after the friendship itself. Borrowing this bond of endurance from Montaigne, distilling it into an approach to others and worlds, allows a construction of friendship as an openness to encounters of not one but many 'true' friends. This very same openness is crucial to beyond-human work: an openness to different relations, space, and temporalities with different kinds of subjects, human and non-human, for beyond-human futures.

Friendships involve shared affection between friends who engage in communication and are mutually benevolent. This bond of friendship can breach barriers of space, time, and difference to share information, knowledge, and support (Bell and Zaheer, 2007). Where proximity and closeness help sustain a friendship, these are not necessarily determining factors for endurance, neither is a friendship conceptualised as ending when communication ends. Rather, friendship endures as a continual orientation towards the other (befriended), a bond which cannot be ended even in death, so long as the friend continues to orientate towards it. Friendship is a kind of love that endures, when love is freed from its historically contingent violence (hooks, 2000). This sustenance of friendship can be understood, in a geographical and embodied sense, as 'holding space.' Holding space for and in friendship is here a geographical and embodied way of enunciating a care for others who may not be physically with, but are emotionally with us: not me, but with me. This holding space can continue to form and shape the relationship between self and other and the condition of self beyond its 'actual' limits. The holding space is one of open-endedness and continuity: an essential criteria to maintaining space (and therefore time) as potentiality and possibility (Massey, 2005).

In animal activist histories, the voices of elite white men, partially rooted in The Oxford Group, spanned Britain, the United States of America, and Australia, excluding other people and perspectives. Through the processes and favours of friendship, their citations centred their ideas and stabilised them as founders of animal thinking. This is rooted deeply within their own identities as friends and an academic collective, which can be evidenced most clearly in the *everydayness* of their relationships and communications: a letter between Ryder and Regan about an all-expenses paid transatlantic trip to promote one another's work (1998/9); or letters between Ryder and Singer organising phone calls to discuss their lives, and their professional futures as intertwined (1973). The kind of friendship philosophised by Derrida (2005 [1984]), Montaigne (2004 [1580]), Nietzsche (1996 [1878]), and Foucault (1997) is reminiscent of this example: a friendship from above articulating itself as resistance to power even as it reproduces these same networks of power distancing from the goods they proclaim to care for.

Contrastingly, radical friendship as 'a way of life' (Foucault, 1997) has always been the case for those on the margins as the very basis on which communities are founded: through an assumed intimacy of being-in-common based upon shared identities, characteristics, or experiences (Kirsch, 2005). Through somatic and proximal closeness, friendships have previously been realised in ideologies of sisterhood (Rose and Roades, 1987), notably in the second wave feminist women's movement, making possible other, radical friendships outside the heteroromantic and familial. Friendship as resistance might be constructed as an attempt to form a different kind of relation outside of power hierarchies and domination, through which a less violent and more cooperative world is possible.

Friendship breaches binaries of public and personal, private and politically, and within friendship, the historical contingencies of reality, truth, and the world are created, constructed, and performed (Webb, 2003): as friends, we become-in-common. As such, it is an absent presence within geographical research and thinking (Hall and Jayne, 2016). This absent presence ignores the socialities and spatialities of friendship necessary to intentional remembering that disrupts dominant histories, where 'friendship as a concept can [also] help to nourish understanding of the complex geographies of human lives' (Bunnell et al. 2012).

Friendship within the archive

The philosophical traditions of friendship think across politics, morality, and intersubjectivity to position friendship as a particular kind of love (Badiou, 2009). Across disparate philosophies of friendship, recurrent themes of mutual care, intimacy, and shared activity (Helm, 2017) and the individual, social, and moral value of these bonds are persistent. For Badhwar (2003), the love of a friendship is one which strives for a psychological understanding and kinship, sparking a mirroring effect to understand oneself better and as a (non-instrumental) affirmation and bestowal of value (ibid., 44), rejecting a traditional Aristotelian conception of friendship as serving (only) the purposes of pleasure, utility, or virtue.

The fluidity of friendship as somehow beyond value, or at least outside of the kinds of economic, social, and political determinants of worth (Perlman, 2018), is made possible by the freedom (which shares an etymological root with friendship) to navigate across spatio-temporal contractions of proximal and somatic distance. This necessary openness of friendship is implicated in a care beyond oneself and a desire for relationality, which remain spatially and historically contingent. In relation to revolutionary politics and transformative activism, friendship is part of a dependency state of relations in which the impossible becomes possible, rather than reproducing desires for independence and individuality that ultimately end transformative possibility (Perlman, 2018).

What is it in the archives that produces a friendship? The feelings of and towards the contents of the archive invoke alternately humour, frustration, and care. Humour is found in activist tactics over the last half-century, publishing under pseudonyms falsifying names and playfully taunting hunt leaders by confusing and cancelling hunts. In blockading and preventing entry to animal-testing labs by gluing locks. In the friendships, teasing and supporting one another. Frustration is encountered when imagining opportunities missed by activists who didn't have the capacity or maybe the will to collaborate with others' actions and approaches. And with care, this history seeps into and breaches the present. Oriented from and towards the past, friendship in the archives is multi-directional, fluid and extended to both humans and animals, even found in the context of these white masculinised histories that dominate. Friendship's praxis embodies virtue, utility, and pleasure and is imbued with meaning beyond myself and beyond the present.

Friendship might also offer a mode out of thinking only with and about animal pain, and into thinking about the possibilities of a world beyond this, where animals are mutual co-constitutors in place and interspecies relationships. The power of friendship not only as sustenance for activism, but as an exemplar of possibilities beyond a present of pain, was looking back from the archives. As I carefully sorted through boxes of papers, drafts, and letters, it was always a particular joy to come upon a collection of some kind: magazines, badges, videos. In particular, Ryder's archives of animal publications across the latter 21st Century offered an insight into the expansive and multiple pasts of animal activism: *The Liberator*, the bimonthly campaigning newspaper of the British Union for the Abolition of Vivisection 1981–1993; *BLACK BEAST* which later became *Turning Point*, an animal rights magazine that ran from 1984; *Arkangel*, a magazine providing animal rights news which ran from 1989 to 1994 and was produced by Vivien Smith and Ronnie Lee, beginning while the latter was in prison. These archives of publications that Ryder held in his archive were beyond his own relatively centrist and academic interests.

Browsing these journals, there was a large collection of the *Annual Pictorial Review* by *The Scottish Society for the Prevention of Cruelty to Animals*, spanning the 1970s and 1980s. The review was themed each year: Conservation Year in 1970, Animal Welfare Year in 1976, Milestones in March 1981. The annual review contained articles and statistics on animals used in experimentation, on legal proposed and passed changes, on public attitudes and media coverage of animal issues, and opinion pieces from activists and pro-animal scientists on animal cruelty. In one issue, there was a 1970 parliamentary discussion on the deafening of birds in a Cambridge laboratory in 1969, which attempted to discover how birds learn to sing (SSPCA, 1970, 27), 'Lord Stonham, Minister of State, Home Office, said: *It would be wrong to think that the experiments had been performed out of idle curiosity, or that the human race was not better off because they had been performed.* It was left to Lord Somers to stress that the answer to this research – whether birds learnt to sing through hearing, or whether this ability was instinctive – was already known.' This case was reported in the context of a break-in to the laboratory where the chaffinches were deadened, releasing them, some budgerigars, and some doves.

My heart somewhere near my throat now, my nerves prepared for pain emerging within the archives in visceral encounters with these pasts, over the page was an insert in the book called *Animal Friendships*. The section ran for around ten pages, and featured likely and unlikely pairings: 'Satu, a three-week-old lion cub, rejected by its [sic] mother, is cared for by Enoch, an eight-month old chimpanzee,' Satu resting his head in Enoch's lap and Enoch wearing a t-shirt (1986); 'PRINCE, the black Labrador, befriended an injured CROW, and they are now inseparable companions,' where Prince and the crow rest their heads against one another (1971); 'Osbert, the duckling, all of five weeks old, felt obliged to take to the water, and involved Chutze, the kitten, who is a whole week older' (1971) accompanies two images of a wet kitten and duckling, and the same pair wrapped in towels. Other pairs include an owl and an Alsatian-Labrador, and a rabbit and a lamb (1970); Ginger the monkey and the cows on a farm in Sussex, and a lion-cub and lamb who were both six weeks old (also 1970); a baboon and fox, who were both orphaned, and a rhinoceros and an Irish terrier (1969). While the anthropomorphism and linguistic objectification reproduces welfarist tropes and politics, these images allowed an escape from pain; tracing the images with my hands and taking them in, the joy and potential of another world – other possibilities and multispecies future – felt closer than it ever had, read through these friendships across species enabled by caring multispecies spaces.

Friendship within archival documentations disrupted the dominant trajectories of pain and suffering, where these are often assumed as inevitable to human-animal closeness (Haraway, 2003). Where there is a failure of imagination beyond pain and domination, friendship offers an alternative. Encounters with archival friends were encounters with life and death, in and beyond a trajectory of friendship as hope and care beyond the archive holds the possibility for becoming-in-common even as it holds the dead. This composition of life, death, and other absent present human and animal subjects combined with an ethico-political pulse of living (and dying) less violently are a sort of 'carrying-on' (Emmerson, 2019), in the form of an affirmative collective remembrance. This carrying-with and moving-on break with the past even as these historical legacies sustain it.

In an archive of animal pain and violence, finding this ongoing series of 'unlikely animal friendships' represented a possibility beyond this history of pain. Being secreted in to the once private lives of archived people and archived events is accompanied with a sensuous and affective pull (Cifor, 2016), not only experiencing these pulses, but deeply implicated in these relations, positioning archivists as witnesses implicated in the ethics and politics of the archive, both particular and general. There must be a certain resistance to the extent in which we talk of archives as lost, secret, or elusive, rather thinking with the classed, imperfect remembrance they offer.

In resistance, a friend does not blindly support, but critiques in striving for something better for friend and befriended and is possible even across the temporal distance of history. This 'archival bond' is an originary, necessary, and determined connection between archival records stored in institutions (Duranti, 1997) so as to unfold and trace documents and materials alongside one another, as well as to ensure records maintain their very existence. Within this bonding, there is an elusive extra-archival bond(ing) that is more spatio-temporally expansive.

In striving beyond the archive, the history itself encounters others: human, animal; alive, dead; object, subject; being, thing. This orientation can also be a disorientation, often overlooked as we direct space intentionally on how things are and where they might go. Disorientation was commonplace in this archive: from the entrance through the messy uncatalogued papers, seeping into the reflective and analytical work, whereby 'the orienting relations to other bodies, to actions and to situations' are lost through 'incomprehension, confusion, and disintegration' (Bissell and Gorman-Murray, 2019, 707–8). Relations, knowledge, and history are undone in a geographical sensitivity both to and beyond this (human) body across elongated spatio-temporal lineages. This is brought further into confusion when the archive was actually breached, and friendships with the historical figures I encountered in the archive become real.

Friendship beyond the archive

During my research in the archives, I had the opportunity to speak with and meet several animal activist-archivists across Britain, who have dedicated their time and physical space to collecting and storing the histories of the animal movement. They are kept anonymous here, but the relationships with them and their collections allow these pasts to live into the present and understand how these connections and friendships continue to shape the movement today. These collectors shared with me objects and histories including the first produced faux fur coat; photographs across the latter half of the 20th Century's animal protests; and books, papers, and correspondence between notable activists and thinkers. As fascinating as the collections are, and as important as their preserving is, equally as vital are the commitment and passion of those collecting and protecting these histories, to ensure the works of thousands of people are not lost but can be passed onto future generations. The enchantment of these animal archives is in their materiality, but even more so in the enchanting relationships of activists with their own history.

On one trip, I walked with my host one evening after a day working through his archives. Walking along the winding roads that evening, we bump into a woman walking her dog. They have met before, but he has forgotten her name. She was and is a vegan activist, recently moved to the area. The grass on our left is brown-green and dry with late summer overgrowth. This woman tells us she is heading out to lay out some water for the badgers who have been struggling in the summer heat, asks if we would like to walk with her. Leaving her dog at home, we walk along thorny paths, tracing the route of the badger sett, perhaps hoping to catch a glimpse of our shy and scary badger friends. Stories of who they know, mutual friends, places, and times they have crossed paths before in the last 20 years blow on the wind back to me. Friendship: it welcomes, it sustains, it overflows.

While we may seek to predict or assume a friendship at some level based on a shared interest in a particular ethic, politic, or friendship, an initial connecting conversation and belief, the spontaneity and unexpectedness of friendship depend on something quite different, something just out of reach of the speakable, something beyond enunciation. My friendship and work within the archive created the circumstances for this friendship to emerge and continue, but it was not only our friendship that mattered. There is an enduring and expansive network: me-archive-archivist-activist-badger are each implicated in an interspecies care for a collective world beyond this space we shared together, realised in our activism and work, in the objects we collect, in the networks we are attracted to, all ultimately fuelled beyond ourselves and beyond the human.

Friendship cannot be deconstructed, reconstructed, or predicted according to a positivist model, balancing properties against one another (Keller, 2000). The ephemeral or enduring nature of friendship escapes this until it is considered spatially, temporally, and historically. This friendship can be thought of as neither random nor predestined, but emergent from particular spatio-temporal gatherings in which those of us who care about animals find one another. Here we bond, imagine, and create new worlds that once opened become ever-expansive networks across space and time. Like atmospheric things (Mccormack, 2014), similar entities and beings are drawn towards similar spaces and actions, which in some inevitability creates seemingly innocuous encounters that was always possible because each friend continually orients towards this network. Or, when a person orients towards a movement – such as animal activism – similar movements occur towards this friendship-network of animals and activists, which in turn envelopes within it the potential conditions for friendships that are ethically and politically motivated to be enacted.

These assemblages of activists inevitably bring people and animals into contact with one or other of these networks. These networks are connected themselves by mobile actors moving between and across spaces of activism, be they academic, organisational, or radical, the same movements spoke of in the previous chapter. This mobility within and through networks of vegans has of course been enhanced by digital media's connective role, but it has also been the basis of a continued and sustained movement and community for animals. It is not chance bringing together and motivating friendships, but a navigation of already constructed networks that remain open to new vegans, constituting an ever-expansive possibilities for transformation implicated in these finite unfinished histories beyond the human and beyond the present. There can be (sometimes simultaneous) deliberate and indeliberate inclusions and exclusions of these gatherings, which in turn can open the network to duplicitous intentions of befriending.

The Special Demonstration Squad (SDS) was founded in 1968 after anti-Vietnam war demonstrations in London. It was initially funded by the Home Office (Taylor, 2015) and in 2014 as part of her statement on the Stephen Lawrence Independent Review, then Home Secretary Theresa May announced a review to establish the links between them. Their covert and illegal activities since 1968 mean questions must be asked of who is being policed and why, but also what the afterlives of those individuals and communities affected are. In 2018, Lush ran a campaign in their storefronts and online that drew attention to the SDS and the stories of the women who were tricked into relationships with undercover police (e.g. Dancey-Downs, 2018) and to those who still don't know the true identities of their former friends.

Animal rights activist 'Jacqui' was the target of undercover police 'Bob Robinson' and gave birth to his son before Bob disappeared when their child was two years old. She was eventually paid £425,000 compensation, but if she 'had a choice – less money and more truth' (Casciani, 2014) would have been a better attempt at rectifying the emotional damages. It was only through the openness and trust of animal activists that infiltrators accessed the community through gendered violences, whereby women were viewed as easy legitimising targets to gain access to the movement. The exploitation of an ethico-political friendship was co-opted and manipulated, violating the possibility of continuation of these groups, and beginning the break with the past evident in contemporary activism.

Undercover policing of activist movements in the United Kingdom has been in and out of the public eye for the last half-century at least. In animal activism, the stories of the women targeted by undercover police, their testimonies and the consequences of this politicised undercover policing are notorious, disrupting long-established networks that appear fluid, but need to be determinedly open to newcomers to expand. After such a violation, there is little hope of continuing with the same people in the same ways in the same spaces, because the bonds built are irreparably broken. The echo of the damages, dangers, and deployments of befriending by powerful actors as a destructive, manipulative, and untrustworthy force must also be considered in any theorizing or practice of friendship. It is not only the power differential, but the potential pay-off of faux-befriending that is violating.

Within the vegan community, there are concerns and cautious beliefs towards outsiders or those new to the network and these have been historically and contemporaneously validated through betrayal and espionage, as well as through more subtle disappointments in potential new friends in the network. In friendship as a gathering, there is a departing from a puritan idea of friendship as between two, instead conceptualising friendship as multiple and open, enduring even beyond death.

Messing (with) friendships

Duplicitous friendship may be thought of in terms of a doubleness, or 'perpetual nextness' (Cavell, 1992, in Hodgson, 2016) towards a desired position of power or esteem. Enveloped in this perpetual nextness is a perceived improvement to their life, but it is unable to progress because ultimately their destruction of the bond of friendship will be revealed. A transformation of the future requires an understanding of the present, its history, and the contingencies determine the boundaries of possibility through the physical, temporal, and spatial ends of the past and future emerging from this present. An open and fluid friendship that remains future-oriented cannot also be inhabiting already and persistently a nextness that abandons its own past. A break from the past is not an erasure of history, but rather a declaration of reorientation.

These breaks, corruptions, fractures, and dissembling of friendship are vital to understanding the shape of veganism. Rather than an attempt at a politically neutral construction of friendship, there is an attempt to understand friendship that does have contingencies and that can be spatially and temporally bounded as and when the subjects of that friendship transform. Within this politically and ethically motivated friendship, the moving beyond of particular temporal and spatial elements of said friendship does not end the holding (of) space for that friendship, nor end the friendship. Rather, in understanding subjective identities as fluctuating and transient, there must be space for different subjectivities of individuals to be in relation. As such, there is not an unbreakable bond of friendship with every face of the friend, but certainly with some (Hall and Jayne, 2016).

Opening friendship to these spatial, temporal, and emotional negotiations and navigations may help us to understand the parts of friendship that do endure and shape activist and revolutionary ethico-political movements. Which is to say, that where friendships include and exclude, bond particular people at particular times and spaces around causes, the actual dimensions of a friendship may be more or less visible at times when these subjects gather. There is also a passive dimension to these ethico-political friendships in sustaining their bonds of endurance. A friendship remains open to not only those subjects but to all who seek memories, experiences, and connection to the wider activist network. This passive yet continual pulse of friendship in vegan activism is not so easily archived but is the foundation upon which the movement itself relies. I contend that this is a feminist cornerstone of the movement that values stories, care, and openness rather than hierarchy, rationality, and legitimation by outside forces.

These networks of caring for animals are approaches of friendship that constitute veganism as a community or movement with bonds to one another and also with particular boundaries. Bonds between friends endure, they overflow life itself returning through imaginaries or impositions upon everyday occurrences after the friend has departed and these may be negative as well as positive. Friendship within vegan activism departs from a singular friendship as truth, or a fraternal exertion and maintenance of power, rather considering friendship as a way of life (Foucault, 1997) for those of us working within conditions of openness.

Perhaps because of the necessary partiality of friendship being in friction with some kinds of theorising that seek impartiality (especially those of utilitarianism or deontology, Keller, 2000; Stroud, 2006), conceptualising and defining friendship as ethico-political resistance remains difficult. Eschewing a 'properties' view of friendship (Stroud, 2006, 167), away from a countable relation and towards one of embodied knowing, is more easily compatible with a feminist ethic, politic, and research practice. Friendship as an approach is a movement beyond oneself and beyond the present, both historically and future-oriented. Expanding and keeping open this approach of friendship is to, again, embrace uncertainty as a principle (Wadiwel, 2016). Such an approach is risky and open to exploitation, but also essential to the reformation of these bonds found within and outwith the histories of animal activism, ever in motion, reconstruction and expansion. This negotiation or calibration 'demands that in the contingencies and vicissitudes of life we possess the insight and the character to achieve truly the good of another' (Sokolowski, 2002, 462), centring friendship as a care beyond the self and for the future from a historically and socio-spatially informed position.

Friendship's social relationships are not constraints, but rather potentials for self and world construction (Cohn and Wilbur, 2003), within which an ethical engagement with friendship itself can emerge. Friendship's fluctuations, orientations, and emergences can be seen here to construct the vegan movement itself as distinct from but emerging out of wider animal activism. The histories can be read as gatherings of people whose friendships preceded and followed their activism: a genealogy reaching for, between, and beyond itself, seeking not to fill silences and omissions, but to recognise and listen to them. This is an ever-open potentiality, which welcomes and excludes insiders and outsiders, dependent on their continued orientation towards or movement away from the gathering. There is not one singular telling of these histories, but there are ones better preserved, stored, ordered, and retold. These are those of the wealthy white male elites whose lives are closer to centres of power.

Summarising Part 1

In Part 1 of this book, I have been situated in the archives of Richard D. Ryder to think about and with the histories of animal activism in Britain. In Chapter 1, I contended that many of these histories reproduce binary thinking and ask how embodied approaches to multispecies worlds allow for a rescaling of the relations between the past and the present. In Chapter 2, I focus on how white, upper/middle-class men have overshadowed the animal activist movement across all typologies of activism, seeking to 'legitimise' human care for animals. In this chapter, Chapter 3, I was concerned with friendship of and in the archives and how friendship itself shapes contemporary and historical networks of animal activism. Taken together, the three chapters form a historical-geographical intervention that critically engages with a long and important political and social movement for animals in Britain. The histories of animal activism in this book serve not as an account, but as a critical engagement with how humans have organised on behalf of other species, but how activism is never separate from the socio-cultural-political milieux in which it is located.

Through a relational spatial lens, these chapters taken together offer a lens into the ways and spaces in which humans have sought to move beyond themselves, and to create conditions for beyond-human care and transformative world-making. A partial history of veganism, activism, and the role of friendship has here been told through these (extra-)archival encounters. Where activist practices take place not only in the upper classes or by white men, there is not the impetus to store and collect the materials, communications, and stories of others in these institutions, because these institutions' own networks are closed to those in the margins. Moreover, the continued silencing of these histories reaffirms once rational, masculine, legitimising power (Fraiman, 2012) that has demanded to be centred in the narrative of animal progress, when this is of course only one part of a much larger and longer history of caring for and about animals. Where there is power in history, this is 'also about the process and politics of its production' (Ono-George, 2019, np).

Likewise, a history of animal activism cannot be divorced from histories of other social justice struggles, as women animal activists advocated and practiced for their alignment with feminism (Frances Power Cobbe, 1822–1904), pacifism, and LGBTQ+ advocacy (Brigid Brophy, 1929–1995) and continue to do so today from broader theoretical and practical feminist positions (Adams, 2010; Harper, 2010; Gaarder, 2011; Ko and Ko, 2017). Veganism has been and continues to be part of a revolutionary perspective for transformative futures not only in a changing food landscape, but in terms of human responsibilities to other animals (Davis, 2016). This is not, then, a break from a rational history, but rather a return to the interconnected revolutionary intersectional struggle that resists the same structures of violence and oppression exerted on different kinds of bodies categorised and treated as less-than-human (Morin, 2018) that was obfuscated by the rational centring of white masculine Western voices.

At the conclusion of my archival research, I wrote in my field diary: 'the archival animal can only go so far in imagining; the present always lingers as I reach for the past. During my time in the archive, time wasn't spent elsewhere. Becoming lost in papers and histories is important and vital, and it allows me to become critical and positioned. But a history cannot be written. It is not singular nor linear and resists being written. I feel distanced from animals. From what I could or should be doing now.' At first entry, I was secreted into the silence of the archive that whispered of unseen knowledges within its dusty boxes of possibility and I approached as a friend, constantly intrigued and turning towards it, active in the relationship with an archive that could not speak back. Then, the silence became immobilising. After the months spent alone but for the past, the quiet of the archive overwhelmed me. Despite becoming implicated in this history, I had become distanced from the reality of the lives of animals with whom I share the present, becoming traces or hauntings (Mills, 2013) of the always somehow unknown and unreachable.

Part II

Presents

Figure II.1 Chickens rest. Copyright Catherine Oliver, 2017.

4 The contours of contemporary veganism

Contemporary vegan activism encompasses a spectrum of approaches, from extraordinary events of protest and liberation to quiet and consistent 'ways of life' resistance. Contemporary veganism is informed by historical animal activism, shaping visions and shared definitions of different forms of activism. When I raised the question of defining activism in interviews, a historical imaginary of the activist subject was persistent: 'Everyone has this image of the 80's and 90's activist, handcuffed to a tractor, without being there back then. To call myself an activist I would be doing more than just being vegan, but that could be as simple as bringing vegan food to work or inviting non-vegans to events or having an Instagram.' The construction Alf referred to is of extremists who explicitly threaten the way of life of the animal-eating majority. This imagery has persisted as a critique of contemporary veganism that has been adopted across Anglo-American discourses, with federal laws in the United States of America passed in 2006 to prevent interference with animal enterprises (Legal Information Institute, 2019).

At the same time as this imagery persists, the expansion and navigation into virtual space has been important for education, outreach, and activism for vegans, taking on a role as part of an effective umbrella of diversifying modes of activism. The imagery of the ideal activist persists as a figure who, untethered from other responsibilities, dedicates 'his' time and life to the animal cause. This figure is deployed as both self-surveillance and community policing of who is 'doing enough of the right thing' (Craddock, 2019), undermining those who cannot or do not perform activism as an absolute. Jack, who became vegan in the 1990s as a child, is sympathetic with fears of militancy from the inside of contemporary veganism entangled with these extreme figures within veganism: 'I worry that militant veganism will become too radical, and it sounds strange to say that as I would never have thought of that ten years ago. I think veganism went downhill when it went over to the USA and they turned it on its head a bit. Gary Yourofsky[1] has a lot to answer for. I always think of veganism as being a British, Western concept as that defined movement.'

Veganism's 'Britishness' is understood in the context of the legal, political, and social histories of animal activism, a narrative emerging in the contemporary era from the Oxford Group, but with gendered histories of animal activism dating at least to the 19th Century, when women's affinity with animals was derided against men's 'reasonable' activism (Donald, 2019). This activism became less 'respectable' (middle class) in the latter part of the 20th Century due to heavier policing, surveillance, and violence towards animal activists since the 1970s. Policing of animal activism is rooted against the actions of Animal Liberation Front activists, whose decentralised structure 'recognised that that way of operating was the most effective in terms of doing the most action and also avoiding arrest' (Boisseau, 2015, 144). This construction of extremism or terrorism was affirmed by media portrayals of animal activists as violent and dangerous, attacking a 'civilised' British way of life (Pfeffer and Parson, 2015, 129) that threatened to reveal that the nation was, in fact, not one of animal lovers.

These tensions continued to fester in British activism and policing, ushering in the 1990s Spy Cops undercover infiltration policing of animal rights groups, amongst other environmental, religious, and anti-racist activist groups (Woodman, 2018). However, it is only in recent years that veganism has itself been deemed an extremist position, historically reserved for animal rights activists, a slight but important difference in framing. Vegans are more typically spoken of in ridicule, as ascetic, difficult, faddish, and oversensitive and only in the rarest discourse being spoken of as hostile (Cole and Morgan, 2011).

The mainstreaming of veganism has, as well as seeing the espousing of hostility and extremism of vegans, offered new ways of relating and understanding veganism as a tripartite practice for health, animals, and the environment. Because of the anthropocentric emphasis on health and environmentalism, there has been a softening of opinion towards vegans not as outsiders, but reimagined as investing in healthy and sustainable futures. In doing so, a vision of human-centred veganism has been eased into the mainstream, normalising vegan practices and removing productive friction between vegans and non-vegans. Located in the contemporary landscape of vegan activism, this chapter explores the historical legacies and contingencies of animal activism, the spectrum of tactics, and the proliferation and successes of quiet activism approaches tentatively offered in interviews with vegans. I end this chapter by exploring how vegans are connecting across struggles and oppressions not as comparative experiences, but instead as interconnected.

Activist histories in the present

Contemporary veganism has put temporal and spatial distance between itself and earlier animal rights activism, emerging as a different iteration or 'wave' of activism (see Offor, 2020). For Sheila, this has materialised in the definitions of 'activist': 'activist is a loaded word and people expect an activist to be doing something more than just not participating in the consumption of animals ... There are different ways to be an activist and it isn't as loaded as it used to be.' The depoliticising of activism is entangled with a break from history and the capitalisation of both sustainability and wellness industries. This renegotiation of vegan and animal histories forces vegans to adhere to unreachable standards of the ideal activist, performing and navigating their newfound truths as if they had nothing else to care for.

The attempt to break with a white-washed history has opened veganism to a more diverse set of images, but these are untethered to the feminist and social justice origins from which animal activism can be rooted. This break with history, while necessary, also risks losing the marginalised knowledges of previous generations, where the mainstream history has been centred as the *only* one. The institutionalisation of dominant histories and the citation of scholars such as Derrida and Haraway in critical animal studies is a wilful rejection of ethico-political lineages of animal activism thought within critical animal work, one that was present before in Singer, Regan, and Ryder.

Where vegans share narratives of veganism as a return to self, this experience is a personal and embodied experience, which rejects previous knowledges of others' caring for animals as an abstracted notion of justice or rights. In seeking to understand embodied veganism, there is a proliferation of vegans who themselves reject or refuse to engage with longer historical lineages of animal activism because in doing so, they do not have to engage seriously with their past complicity of violence towards animals, instead allowing this to be understood as a failure of societal education.

Asking Amy, a vegan for 18 months and vegetarian since her childhood, about the influence of historical animal activism, she thought for a moment before saying, 'I know that things were acceptable in the past and now they're not so when you give it thought, you sort of realise there has been a progression.' The stalling of intergenerational conversation and the divide between generational vegan and activist approaches was recently described by Roger Yates as a 'battle for ideas' (2020), which has manifested for some vegans, especially those who have been in the movement long-term, as a goal to recapture the radicalism of early vegan philosophy. But, as demonstrated in Part 1 of this book, these leave open questions of whose history, whose philosophy, and what to do with the new forms of embodied ethics and politics emerging in contemporary veganism?

This history of activism is being negotiated by vegans who have not always necessarily looked kindly upon activists coming to terms with how they belong in this landscape and history. For Rob, who has been vegan for two years, his reflections on the history of activism were that 'the image you get is that these are the people breaking into the laboratories, the people that used to stand in the city centre with pictures of animals being abused being confrontational and I've never really thought about the history of it all.' There is a distancing between his own embodied ethics of veganism and those of prior generations of activists whose direct action has become a memorialised construction of animal activists as 'extremists.' At the same time as contemporary veganism is distanced from history, the same stereotypes of vegan activism as only exceptional forms of direct action reinforces activist 'impostor syndrome' (Craddock, 2019). For most vegans, this 'ideal activist' trope is at once out of reach of their imaginations of themselves, but seemingly venerated, making them feel out-of-place in the vegan community.

Vegans have begun to re-define the meaning of activism not only to themselves, but to the people around them; in Rachel's life and networks 'people are very shocked to find out that I am a vegan because I look relatively normal.' Mediated constructions of activist histories, and white middle-class representations of veganism of the present (Morris and Oliver, 2016), fuel not only outsider but insider perceptions of veganism and activism. This perpetuates myths of the ideal activist that are dangerous and in need of deconstructing, rebuilding, and shaping to not be untethered from responsibility, relationships, and the world, but rather accountable beyond activism. In the age of social media activism, these constructions of 'good' and 'bad' activism have proliferated in forms of aesthetic and 'clickbait' activism. In particular, vegan street activism has grown, with urban spatial interventions such as 'Cubes of Truth,' where activists persuade passers-by to watch virtual reality videos of slaughterhouses, and 'Change my Mind' tables, where an activist invites people to sit with them and explore arguments against veganism. The filming of these forms of activism is infinitely 'instagrammable,' curating short videos that can be shared to support veganism's online presence.

As well as participating in the action, sharing the videos online has become a form of quiet activism, which exists in a guilty relation with a desire to be able to engage with these tactics even if not in a physical way. Even as a regular attendee at activist events in-person, Esme shared these emotional challenges, 'I feel guilty because I'm not doing anything apart from sharing stuff, and even though I said that that is activism, I don't know ...' and this sentiment was echoed across interviews as vegans navigate across virtual and direct action, their own comfort in activism and what they feel is expected of them. Performing exceptional forms of activism were understood by some vegans as a process of authenticity for those at the site, but for those who aren't there, these videos are useful to share into their own networks as a form of online activism (discussed in Chapter 5). Rob, for instance, 'would say I am not a traditional activist in the sense that I'm not out on street corners, all sort of evangelical activism if you want to call it that.' Contemporary veganism is negotiating outsider and insider constructions of how an activist should be, with emotional and exclusionary consequences.

Briony, a vegan for three years who is based in the North of England, described how she often felt isolated from both her family and from vegan communities, she 'gets really upset with it all,' especially her feelings that she should attend activist events at the same time as recognising her emotional and mental health would be at stake if she did so. Briony conceded that, if she wasn't geographically distant from any activist groups, 'do you know what I quite like that's not dead annoying and its quite in your face is where they have tablets with videos on, they're not saying anything to people, they're not grabbing people, they're just standing there and people can watch it if they want, I like that.' This quiet intervention of space is, in Briony's imagination, a subtle but effective way to transform public space that brings together the virtual and public elements of activism. However, she has not engaged in this kind of activism.

For Alf, 'times are changing, there are more and more people and there's more information basically and people are sharing that information. I think that is how more and more people are going vegan: they see a post on Facebook, for example, saying meat does this, factory farming is this and it shows that it is wrong,' and these are accompanied by 'documentaries and stuff so that there is more information that is more accessible as well so I think, you don't need people on the streets.' The historical connection of 'on the streets' activism is fading, yet the imaginary of ideal activists persists for vegans, calling themselves into question for doing 'enough.'

Alongside the rise of video and photograph shareable content activism is a return of a discourse of animal rights in contemporary veganism, borne (perhaps unknowingly) out of the histories discussed in Part 1, yet not directly engaged with them. Returning to the pressures of dedicating everything to activism, the emphasis on activist's dedication is gendered and raced in particular ways (Craddock, 2019). With the rise of social media, vegans are building careers from full-time paid activism, and these burdens weigh heavy on those who cannot, and choose not, to do so.

A historical spirit is understood in different ways in the vegan movement; for example, as entangled with Britishness (Jack), anarchist ideologies (Alf and Titan), and wider political beliefs (Esme and Sheila). Yet, there remains a persistent lack of inter-generational conversation, which may be part of a wilful refusal of the knowledge of long histories of veganism that would instil further guilt for vegans not 'knowing' earlier about veganism (McDonald, 2000). Re-establishing these historical lineages could resist, navigate, and re-define what it means to care about and live in less violent multispecies spaces.

This wilfulness in the contemporary vegan movement's ignorance of its pasts divides newer vegans, who centre embodied truth, from older generations of vegan activists, still attendant to rational ethics and their followers seeking to foreground animal rights discourse. Felt connections with animals have not necessarily been absent from earlier waves of activism, but only in recent years has empathy or compassion been recognised as a valid motivation (Stallwood, 2013). This approach remains at risk of co-option legitimising white, Western voices, when it should be rooted in the marginalised groups to whom it belongs and has long been a way of life (Stallwood and McKibbin, 2019). Distancing animals from emotion also distances veganism from its rooting with the women that shaped it (Adams, 1994; Oliver, 2018). The contemporary political turn in animal theory and ethics may yet serve to reinforce rationality over emotion (Donaldson and Kymlicka, 2011; Cochrane et al., 2018), or this political turn may necessarily involve new conceptualisations, bringing in emotion and embodiment as political forces.

Quiet activism, eating and everyday life

Finding a history that may have already been lost heightens the challenges of intergenerational activism that can reinforce 'a sense of perpetual debt to the past' (Malizia, 2014, 93). Without this careful connection between present and past, star systems of activism (Freeman, 2012) can already be seen repeating, upholding white men in ideal activist and saviour roles as the ultimate knowers, distributors, and leaders of animal activism, a mistake repeatedly made in animal activist histories. Contemporary activists are navigating these histories within new landscapes and means of action in the present across various typologies of dissent (see O'Brien et al., 2018). The normalising of a spectrum of activism disrupts notions of the ideal vegan activist to centre the realities of veganism and activism practiced in mundane ways. This is also a rejection of activism as untethered from gendered, raced, classed realities (Ramírez, 2014) particularly through online activism.

Online spaces are used not only for organisation, but also for resistance, education, and outreach. These tactics on and through social media and technologically mediated approaches are vital to open and fluid activism: 'with social media people can participate in activism in a way that feels right to them and still have boundaries' (Alf). Others, like Jack, remain sceptical of the potential to sustain a movement only in online spaces for education, activism, and organisation, who surmised that 'I think all you are doing is slagging people off from the protection of your desk. If you really want to have some impact, do what animal rights activists have done for the past hundred years and go out and physically do something.' Jack's history and involvement in anti-circus campaigns, notorious anti-vivisection campaigns, and vegan-run animal sanctuaries impact how he perceives contemporary activism as less real or effective than campaigns disrupting traditional public space.

Jack claims that social media has seen rise of populist veganism, a 'hippy trendy thing that has changed the philosophical meaning of what it is all about.' He views this as detrimental because he perceives a lack of rooting in the philosophical underpinnings and histories of veganism. In doing so, he is perhaps complicit in upholding the white, masculinised ideal activist, and policing the boundaries of who is welcomed and Jack's rejection of new forms of activism is representative of a felt closure by older activists to newer vegans. This was captured by Matthew who, when we discussed histories of direct action shared that 'understanding the past to articulate and understand the future is really important, but I don't think we should allow that to really divide and fragment the movement, you know, you've got your old school animal rights here, you've got your modern plant-based's here and that's just not helpful.' Matthew recognises the importance of these inter-generational conversations but has felt these more keenly as conflicts that impede veganism.

In upholding real activism as that on the streets, it excludes women, people of colour, and people of different abilities and classes who are traditionally marginalised or marked out in public space (Kern, 2019). Social media activism is accessible to many more people and perceived by those part of the new surge of veganism as, for the most part, positive. Many younger vegans viewed their experiences of the world as mediated anyway by screens. Esme notably was keen to share how social media has allowed her to take part in activist events, build activist networks, but also in ways that are accessible to her and generous to others: 'Some people may not be able to go out and protest, mentally or physically. They might not want to go and do them because it can be really stressing. I think it is important that everyone does their bit in whatever way is good for them. It could be even just posting on Instagram, and it is different for everybody, but as long as people are promoting the message, that is activism.' Esme understands activism looks and is different for every individual, because of how people inhabit the world, their access, and exclusions.

Accounting for different bodily and emotional capacities acknowledges the potentials and turn to transcorporeality within and between members of the community, as well as care for inclusion and of others. The removal of barriers in activism is part of a move away from rationalism and towards a more intersectional approach, prioritising different embodiments of veganism, and variances in navigating social space. In being flexible to work with available means in creative ways, more people can be included in these new worlds, in ways that have not always been possible in white, ableist Western veganism.

When veganism comes into friction with non-vegan spaces and people, it is a refusal to be pushed out or silenced that emerges as important, rather than extraordinary liberation and protest domination narratives. A refiguration of activism as a way of life (Foucault, 1997) relies on understanding wider navigations of the world and space. For Daisy, vegan activism as a way of life is explained as 'I don't think you have to be out there barricading yourself to something. I think it's important for people to talk about this in daily life, for it to be an approachable form of activism ... You are making a protest every time you go to the supermarket and refuse to buy products with animals in them.' This shift from a life dedicated to a singular truth towards lives oriented towards less violent multispecies worlds is a result of vegans' refusal to remain outsider and an opening of the everyday. Daisy politicises the foods and products that she buys (or does not buy) as a form of consumer boycott (Brinkmann, 2004) that disrupts the lives of those close to her, opening space for conversations as activism. Veganism can thus be understood as a sensible response to reduce personal involvement in (capitalist) violence (Gelderloos, 2008). The consequence of these refusals is activism in intimate, everyday ways. Veganism's refusal of some spaces, actions, and beliefs is an opening of other spaces, times, and ways of being.

By seeking new connections and spaces to open conversations about veganism, couple Sheila and Alf understand their vegan food business as occupying deliberately this space of tension by only working at non-vegan events. 'As a direct consequence [of running a vegan business] we get groups come to us for meet ups, buying food from us, organising events while they are there, making friends and networks with other vegans,' Sheila explains. In bringing vegans into non-vegan spaces, they curate and offer a space of resistance within what might otherwise be a space of mourning for the animals being sold as food around them (Stanescu, 2012). Alf adds that 'Running our business, we define as part of an activism, an activist cause through providing food. Our name contains vegan, it is putting a positive message out in the streets of wherever we trade. We get asked a lot of questions about it and we try do it more subtly. Everyone has to eat food so ... that kind of link between vegans and non-vegans so conversations that aren't going to result in slanging matches.' The negotiation of sites of potential tension being co-opted for subtle activisms by Sheila and Alf is also operating on the smaller and closer scale of the familial and friendships.

'I had to sit down with my parents and have a chat about veganism. They were brought up quite old-fashioned and it upset them a bit,' Freya told me. Her parents were concerned over her health and what appeared to be a radical transition to veganism. Even where ultimately family and friends were accepting of veganism, the feeling of impending rejection was present, most presently felt by Rhys who, a couple of days before visiting his parents, who were themselves vegetarian, and couple of months after becoming vegan 'emailed my parents and said just to let you know, I'm vegan now and if you want to have a chat about logistics of meals, let me know. I really put off telling them for a while, I could have told them much earlier and my mum was like that's obviously fine it would have been better if you'd told us with more notice, but I remember putting it off and putting it off because I was so worried about what the reaction would be which is bizarre.' In everyday encounters, conversations and performances of veganism affect and imbue the possibility of transformation in these relationships and, alongside this, is a nervousness that precedes an assumption of defensiveness, whether or not these fears are realised.

Creating new spaces, like Sheila and Alf, works in conjunction with injecting familial and friendship relationships with new frictions, tensions, and possibilities of veganism. This opening is a navigation of the personal, collective, and worldly as historically contingent but oriented towards a more inclusive future of embodied and emotional ethico-political practices of truth. The opening of space offers those different capabilities, responsibilities, and capacities to become part of an activist movement without abandoning the rest of the world.

A quiet activism of 'small, everyday, embodied acts, often of making and creating, that can be either implicitly or explicitly political in nature' (Pottinger, 2017, 215) is present in the habitual practices of veganism. Spaces of vegan food consumption are ones of heightened surveillance and, here, veganism might be conceptualised as quiet activism. Revealing their veganism within the intimate settings of eating became a route for some interviewees to talk about how and why they have chosen to reject the consumption of animals, in the presence of animal corpses. Because the sharing of food is so endemic to cultural and social bonding (Matthee, 2004), some saw these spaces as ones to expose others to veganism. This performance of veganism as refusal simultaneously opens communication and possibility, whereby the consumption of food itself gives off a particular impression and identity.

For Charlie, this entangles class and gender politics: 'If you are a middle-class man, there are certain privileges that afford you to be able to say you're going to be vegan. If you are a working-class man, masculinity has to be preserved in a very different way. You can't say you're not eating meat because it's such an integral part of being a man.' This negotiation of space as a vegan is also troubled for people with already strained intimate and anxious relationships with food, such as for Titan who has 'a very complicated relationship with food since I was a teenager. I don't like eating in public or in front of people, and if someone draws attention to whatever I'm eating, it gives me mad anxiety. When I'm out, if there isn't a vegan option, I won't ask. I don't want to attract attention.'

Not all vegans have the power to loudly declare or call attention to their consumption, mediated through not only bodies but societal- and self-surveillance. These modest acts of refusal and opening are political ones, reaching for 'gentleness, slowness, subtlety and subversion' (Pottinger, 2017, 216), that remain imbued with a political anger and pain inherent to the animal connection in veganism (Oliver, 2020) where work is being done to maintain previous identities alongside veganism. Once the violence of eating animals that is deliberately hidden and distanced is revealed (White, 2015), vegans become hyper-aware of the injustices and violence behind society's damaged relationship with animals, building and renegotiating their identity in flux with their history. Veganism can overwhelm and encompass a person's ability to be at ease in society, yet through quiet activisms, there can be a persistent pulse of engaging and orienting towards animals through everyday encounters.

This maintenance of openness in personal relationships[2] and the power of conversation was prioritised by vegans as an important factor that had influenced their own veganism, influencing in turn their perceptions of quiet activism. This is captured in several quotes, the first from Matthew who has come to understand that 'intimate everyday conversations are a form of activism. They are in many ways more effective because you can challenge people in a way that they may be ready to listen.' Relationships with vegans was a motivating factor for becoming vegan for Rhys who, in turn, now recognises this as part of his own activism: 'what made me become vegan was being in close proximity with vegans, having conversations with them and using that to reflect on my own decisions, not from any campaigns. If a gentle conversation with somebody about veganism can invite them in, then it's activism.' The accessibility of veganism is experienced on both sides for some, like Rhys, who transformed his own practices after living with vegans and witnessing and sharing in the ease and politics of their lives.

Similarly, for Rob, 'quiet activism makes veganism accessible to people, brings it to them instead of people seeking it out.' This sentiment suggests a quiet approach to activism is a revolution of the everyday, rather than reserved for exceptional spaces of performance and protest. Throughout the interviews, the information shared with me – facts, statistic, and stories – has a strong rooting in the ethics and politics of veganism from an educated and active community who view veganism as embodied, but also a movement of justice. Vegan activism is entangled with closeness, connection, and conversation as a new ordinary, within which we might find the extraordinary. This activism is, as summarised by Titan, shaped not by any fixed or static actions or performances, but rather by 'Information sharing, debate, discussion and community. Anyone who goes out of their way to learn more about the way the world works instead of just complaining about it is an activist. Bonus points for suggesting alternatives and spreading the good news.'

The spatio-temporal conditions for beyond-human care

Veganism's intersections with other social justice struggles must find new means of expression outside of essentialism, much like animals' lives shouldn't be judged in anthropocentric frameworks. Nonetheless, comparative politics remain powerful forces. Critiquing the spatialities and structures of oppression and domination offers a less emotionally, historically, and socially volatile conversation over human trauma as a comparator for animals' emotive and embodied experiences of violence.

Shared interspecies oppression has a long history in animal activism – Martin's Act of 1824 successfully used the rhetoric of slavery for animal welfare before the human Slavery Abolition Act 1933 (Kean, 1998) – but 'sceptical audiences are often hostile to expressions that expose animal exploitation through human suffering; they are more willing to acknowledge animal abuse if differentiated humans because human suffering matters more' (Socha, 2013, 223). These reflecting, refracting oppressions, structures, and spatialities must be attended to carefully in order to be integrated into an expanded compassion and an enlargement of the concern with justice (Smith, 1997) beyond essentialist and reductive readings of other's experiences.

When discussing these comparative politics and their emotional implications, Alf reflected, 'every time I have seen collaboration attempted, it results in in-fighting and arguing. People get offended that they are being compared to animals, and this is belittling their experience. As soon as human emotions get involved, people get defensive. People have tried to talk about animal rights and racism, saying the way slaves were treated cannot be compared to animals or the holocaust to animal agriculture because you're comparing humans to animals.' These comparisons are often deployed without attentiveness to how they speak of others' worlds and traumas, as they carelessly reveal structural multispecies violence as if it is individualised (White, 2015) revealing the need to prioritise 'strong advocates for other animals who recognize the structural basis of the oppression they oppose' (ibid., 254).

Becoming vegan is at once a return and a reach; overflowing with histories entangled with this moment and searching and striving for a transformed future. Veganism is a disturbing and disturbed navigation of the world that creates and experiences friction from society and space. The embodiment of veganism is attended to in the following chapter, where tangled conceptions and conflicts of what vegans describe as their 'truth' are explored. In moving forwards to understand veganism, I close this chapter with a lengthy quote from Titan, whose insights echo through the following chapter:

It is impossible to talk about veganism without drawing it into a billion other discourses. Not only is collaboration possible but it's vital. People are so desensitised to social issues that we must provide a unified voice. Consider, you have an in-depth discussion with someone about the importance of veganism for hours, maybe over the course of several days. At the end of it, they go home, and they have a lot to think about. They find some arguments for and against what you've talked about. Mostly against. Maybe they change their mind or maybe they don't. The alternative: you have a discussion about the inter-related issues of

modern culture and society. You talk about veganism, environmentalism, capitalism, feminism, racism. They go home and they're browsing
articles, discussions, free-source academic articles. Maybe they start a
discussion with their friends. They can't stop noticing how that thing
you said about that injustice kind of applies here too. And there. And
suddenly, the world is new and exciting and terrible.

Notes

1 Gary Yourofsky is a controversial figure, who is renowned for encouraging
   animal activists to use violent comparisons of human and animal oppression,
   that attack non-vegan women in particular.
2 My friend, Titan, spoke on this in our lives: 'I met you, who was a fierce feminist and vegan. You were doing your Masters' degree on veganism at the time
   and the more we swapped ideas and the more I learnt about it, the clearer it
   became that becoming vegan was ethically something I fundamentally connected with.'

5 Embodied knowledges, 'truth,' and veganism

Veganism often involves a sense of bodily and worldly 'wrongness,' informed by encounters with violence and pain caused by eating animals. This affects navigations of the world in subtle and explicit ways. When I asked activist and vegan business owner Sheila about her changing relationship with animals, she told me: 'if I saw an animal, like a bird, who has been run over, it would make me sad. Even though it is dead already and I didn't know the bird and never saw it, I feel pain when I see animals that are in pain or who have suffered ... I put myself in that animal's position and I think about their connections in their own life, so you know if an animal has been run over, is another animal wondering why they never came home?' Her reaction to animal suffering, even when pain is not consciously inflicted, is a pain felt in her own body because of a world transformed by and for humans. This story bears resemblance to Richard Ryder's (1998) own account of a similar encounter with a dead blackbird igniting his interest in animal pain. In the first section of this chapter, this discomfort and 'wrongness' is explored as encompassing variously a sense of injustice, a disturbed sense of place, and distressing (bodily) implications of pain and death.

Pain can breach human-animal somatic distance through elongated historical and contemporary orientations, changing attitudes and behaviours towards animals. Encounters with animal death seep through everyday spaces, and sensitivity to this spatial violence is commonplace for vegans. This can lead to a form of dissociation with human social worlds and a retraction from spaces of pain and death across professional, social, and personal realms, as well as a breaking of relationships with non-vegans. The corpses of animals particularly force a confrontation with their own complicity in making animals into corpses. In the corpse, vegans witness the truth of suffering in this corporeal embodiment of pain in death; the body exists beyond itself as a contingent and shifting being-thing, as discussed in Chapter 1.

Where life always flourishes together (Ingold, 2011), death also implicates beyond the dead, but life and death are not offered equally to even all humans. A process of distancing is present when thinking about animal lives and deaths, dictated by their species, location, and closeness to humans. Veganism disrupts the categorisations of animals we love and animals we eat, with moments of horror and disgust at the realisation these animals are one and the same. Amy shared her thoughts on the ranking of animals, 'even when I was a vegetarian, in the back of my mind I had that there are some animals that we take things from and other animals that we don't. In your mind it is easy to make these distinctions but since going vegan, I see [animals] in the same light. They just want to be and be left alone like any living things does.'

Similarly, Rachel, a schoolteacher and vegan who initially went plant-based for health reasons but with increasing research into animal agriculture had her practices solidified as vegan for animal and environmental reasons, told me of the emotional toll of her transition: 'When I first had that revelation moment, it was quite difficult. I cried at random stuff. I was reading and reading and reading. I was listening to [documentary] "Farmageddon" when I was driving to work, and I'd have to pull over and be like "oh my god the bees! The bees!" as fresh waves of realisation and fresh waves of horror came over me.' Vegan knowledges have the potential to disrupt and destroy what came before, manifesting as a desire to break from pasts of complicity and move towards a creative constitution of 'new' futures. Encounters with animal pain shift knowledges and senses of place and self in this transition towards veganism that is defined by process of self-education and renegotiation of ethics, practices, relationships, and space itself.

When vegans learn how animals are treated, they want to end their complicity by becoming vegan and beginning anew, but the absoluteness of animal pain upon which ordinary life is built continues to disrupt their world. In Amy and Rachel's encounters above, the time between these two stages is disorienting. After before and before after becoming vegan, there is an ongoing present of uncertainty and transformation. This ever-present can feel as if it breaches time and space, stretching into an ongoing moment of learning through encounters with what vegans call their 'truth.' 'Fresh waves of realisation' are constituted in engagements with the world and this process of re-orientation is neither painless nor immediate. Rather, it manifests in a simultaneous distancing and closeness across personal, collective, and worldly spheres that reveals what once were solid boundaries as semi-permeable. In this chapter, I share the stories of vegans in Britain, their relationships with 'truth' before, during, and after veganism in their everyday lives and spaces. Attending to navigations of their relationships with themselves, their friends, and family, and the world itself, where veganism becomes a narrative for dealing with their own past violence, transforming themselves as an embodied and spatial practice.

Eating animals, navigating distance

The body's semi-permeable boundaries are revealed when we eat. Philosopher Annemarie Mol (2008) writes that when she eats an apple, or rather this apple, political meaning is written upon and within the apple she does (or does not) eat. The apple she eats is evocative of violence, 'imported from Chile, and thus stained with the blood spilled by Pinochet and his men' (ibid., 29). This apple is inscribed with meaning outlasting the political situation of Chile. Mol's distaste at the Granny Smith is a breach of categorisation – the political felt in the body – and the temporal: a history in the present made possible through this embodiment. We eat not only what we eat, but also consume the process of our food. We are, as such, complicit in who we consume, and how we came to consume them. We are not what we eat, but who we are when we eat is not distinct from who we are, which is to say, if violence is knowingly and avoidably consumed, the eater becomes and is violent. The body is not under the control of the eating subject: what is absorbed and what passes through cannot be mastered, forcing the eater to relinquish control, never wholly subject within or beyond the self.

Confrontations with the violence of who was eaten before veganism force emotional encounters with one's own past. Esme, who is involved in the SAVE movement, participates in vigils held at slaughterhouses (see Lockwood, 2018). She said, 'a lot of new vegans go to solidify their beliefs because it is easy to say, hypothetically, you'll be vegan forever. I think a lot of people go [to SAVEs] and can say this is absolutely the way I am going to live my life [as a vegan] because you come into contact with the victims.' Facing who she has eaten is important to Esme in solidifying her beliefs and practices. Meeting animals in these contact zones near to death is a necessary confrontation with a past complicit self. Travelling to these sites is not easy, practically and emotionally, and activists tend to carpool to these demonstrations, meeting through Facebook groups or travelling with friends.

The unknowing of animal slaughter is enabled by deliberate constructions of space that distance the slaughterhouse, battery farms, and other carceral institutions of animal eating to the margins of society (White, 2015). The institutions of the animal-agricultural industrial complex resemble uncannily the carceral-industrial complex from an aerial view (Morin, 2018), as a 'place that is no-place' (Villaes, in Fitzgerald, 2010, 60). Lacking identifiable structure, slaughterhouses look like prisons, schools, and hospitals; their distance from wealthy residences and the carceral logics maintaining their prisoners as less-than-human separate life from death inside and outside the institution.

The SAVE movement centres interspecies empathy by seeking recognition and reciprocity through closeness. At demonstrations, activists hold vigil and offer water and comfort to animals through the crates of trucks, live streaming to their social media networks. Becoming proximally close to animals and specifically to animal pain and death, Esme's vegan beliefs and practices are affirmed in the undeniable feeling of wrongness in this space, extending beyond the silence of the vigil. Sharing via live streaming serves two purposes in this space: authenticity and accountability. Authenticity in sharing this pain solidifies the belief that 'seeing into their eyes' forms a bond before death that extends beyond it. This encounter makes certain beyond doubt this death's wrongness (Wadiwel, 2016). Proximal closeness opens the possibility for imagining somatic closeness, creating authenticity to speak on these animals' behalf. Accountability to animals with whom a bond has been formed is also a performance of accountability to themselves to share this knowledge of pain. These encounters with bodily pain and death can be recalled and relied upon in future doubts, to solidify a break from a past of eating animals and ensuring vegan futures, and the encounters also serve vegans who cannot physically attend activism with animals to negotiate their own relationships with animals and perform their veganism.

Becoming proximally and somatically close to animal pain and death also has a voyeuristic element, serving human social relations and performances of veganism that can both build and break bonds. Briony often feels her veganism isolates her within familial, collegiate, and friendship relations, 'I wouldn't try and make life harder than it needs to be.' She feels unable to participate in activism because she lives in a rural area and feels outside of these communities, but she did talk about how she consumes and participates in activism through live streams: 'I don't know anybody and I'm not really one to just turn up to things. I'd have to be took along and introduced. That's my personality. I can't say I've ever been reached out to do anything, and I don't know if I'd say yes if someone did … I've said I'd like to go along to things but then I get really upset. I'd like to go to those pig saves where they witness them going into slaughterhouses, but I know it'd be awful, and I don't think I could do it.'

Briony shares videos on her social media, hoping to continually share information in a strategy that activists use across social media to make human cruelty and killing of animals ever-present and, even if scrolled past, videos and images will keep reappearing on timelines and breaching into everyday spaces of the online. When she watches these videos, Briony moves closer to a networked veganism, but never quite feels part of it because her interaction is mediated through screens. She compares herself to ideal activist vegan tropes, that glorify activists as untethered from life's responsibilities (Craddock, 2019). The potential of social media in activism emerged early in all the interviews and is a vital medium for contemporary veganism that is rooted in longer histories of (undercover) exposure of animal pain and death. These strategies no longer rely on traditional forms of media to pick up their stories but build their own networks as influencers on platforms like Facebook, Twitter, and Instagram.

Present and past activist lineages coalesce in attempts to enact embodied reactions through mediated technology, trying to elicit conversions to veganism. This virtual encounter of human and animal in activism is important for understanding how 'truth-sharing' works to construct a narrative of return in veganism, whereby a new, 'better' self can be found in veganism.

Embodying 'truth'

Proximity is important to creating the conditions necessary for recognising shared embodied worlds and, here, the importance of bodily knowing is foregrounded where rationality falls short. This embodied knowing is conceptualised through a theory in and of the flesh (Moraga and Anzaldúa, 1981), borne out of the physical realities of pain. This fleshy reality is entangled with our ways of knowing where single bodies and selves are always multiple. These relations re-orient us in social, political, and cultural processes contingent upon our singular-plural position. A theory in and of the flesh read through pain is felt as an inviolable bodily and emotional knowledge of the world (Scarry, 1985). To betray that knowledge would be to betray oneself. Acquiring this body-knowledge disturbs a strange same-but-different world.

For Rachel, it was an embodied reaction to knowledge of dairy cows' pain that solidified her (ethico-political) veganism. She originally began following a plant-based diet for a stomach issue. In researching her condition, Rachel came across information about the dairy industry which 'horrified' her and so continued to research animal agriculture and transitioned to veganism. This transition was not only dietary, but an ethical, political, and linguistic transition, moving to a position 'absolutely about the animals.' Rachel imagines herself somatically closer to animals through bodily pain. It was this embodied knowledge that solidified her transition to veganism: 'I call myself a feminist and yet contribute to a system that impregnates and abuses female animals and treats them purely as reproductive organs. I wouldn't like anybody to do that to me. There is something about the prolonged suffering of dairy cow who day after day is going through this.' Rachel became closer to animals through shared bodily ways of being in the world and, through this somatic closeness, felt more able to build a web of connections in which to locate herself and her veganism.

This sentiment was reiterated across interviewees, understanding this embodied connection between human and animal as the definitive impetus of veganism that enables all other. This is nurtured by vegans as 'the truth.' Because of its origination as a bodily knowledge, this truth's power is in its bodily situatedness, felt solely as their body's own. Embodied knowing is a powerful and inviolable manifestation. a gut feeling that cannot be rationalised. This embodied knowledge can overwhelm in the face of ultimate truth, in is 'an instant "sensing", a quick perception arrived at without conscious reasoning... the one possessing this sensitivity is excruciatingly alive to the world ... This sense is developed when 'we have all sorts of oppressions coming at us, we are forced to develop this faculty so that we'll know when the next person is going to slap us or lock us away ... it's a kind of survival tactic that people, caught between worlds, unknowingly cultivate' (Anzaldúa, 1987, 38–39).

This embodied renegotiating of the world through perceived and felt threats to the body is perhaps found in veganism as a secondary *embodied sensing,* one that perceives the threat to other beings' bodies – animals close now to the self – as a threat to their own bodies and worlds. The worlds under threat are performative and multiple dynamic constructions created in socio-spatial relation as actual, real, and potential. When striving for perfection, the body and self are constructed in three ways: the body (and thus self) is of the past, in need of transformation; the body/self-in-progress of the present, doing work to be 'better'; and the body/self we imagine in the future, after transformation (Widdows, 2018).

This future is not a manifestation of others' societal beauty and goodness, but rather a projection of ourselves in the future. All three of these selves are contained within an ever-present of transformation. We inhabit these three selves, but each of these remains endlessly multiple (Butler, 1990). These constructions of embodied selves are shaping expansive vegan worlds, being constructed and experienced through these imagined, actual, and real selves; the world we want to be in and the world we are in are differently located spatio-temporal visions. This can be understood beyond linear temporalities, whereby the worlds we are in are pasts, leading right up to this moment of the present, and the worlds we desire are the futures sought through less violent practices. We are caught between the two in an ever-present transformation, where the present has multiple entries.

For Shane, his narrative of veganism allowed him to construct a 'truth' of the world that he says he was always aware of, but deliberately turned away from so as not to change. 'I kind of knew stuff went on but I wasn't aware of the reality of it for a lot of the animals or all of the animals consumed, and I think like I said once I found out about it ... have a lot of faith in humanity, I think generally most of us are good people and I think most people would see that and say like yeah man that's bad man, that's horrible ... and change.' These transformative moments in vegans' beliefs and practices reflect the multiple, ongoing worldly, worlding transformation that is being refracted through the vegan community as individuated and collective, through personal commitments and practices of veganism.

Individual journeys of veganism occur within open networks of possibility and it is within these possibilities that future worlds are enacted. If we have always and will always remain in this between space, just beyond the world of the past pain-causing but never able to reach the desired non-violent world, then we must have inexhaustible potential for desiring transformation. In transforming worlds, this desire pervades and persists throughout revolutions of body, mind, and world. These cyclical repetitions of space, time, and being maintain the potentiality and actuality of worlds that can and have changed, that embrace progress, difference, and creativity of emergent worlds. How do we bound, then, the worlds and selves and commit to an ever-present of ethico-political transformation whilst critically reflecting on veganism's construction of 'the truth'?

Performing the truth

If 'it is questions with no answers that set the limits of human possibilities, describe the boundaries of human existence' (Kundera, 1985, 135), then boundaries blur between beginning and end, self and other, being and thing. These embodied confrontations with truth force performances in the present to be ones representing a body of lightness and virtue. For Charlie, these contradictions between his body and other's perceptions of his vegan body are often conflicting between his ethical 'feeling good' and his physical health, 'I'm also doing this for how my body feels, my body's comfort. I feel like there is this body justification. But at the same time, I feel like I need to say that I miss things to not position myself as this kind of all-knowing person who's doing the "right" thing all the time and is not conflicted about it. I have to take up these multiple positions, so I say I do miss stuff sometimes, and these aren't lies, but it is part of a performance.' Understanding veganism's performance reveals a continuing negotiation in and of relationships. In feeling questioned and questioning his own veganism, Charlie doesn't doubt his truth, but doubts his ability to convey it.

Although veganism's 'truth' is re-affirmed through Charlie's performance of it, it eventually becomes a mostly banal and mundane habit. The intention and reception of action are blurred (Goffman, 1956), meaning performances around food are continually being renegotiated in different spatial and social locations. Vegan consumption is under constant self-surveillance and the surveillance of others, meaning that even when there is an appearance of fluidity, there remains a disjuncture and discomfort of veganism in a non-vegan world. Vegan performances are heightened when the presence of a vegan disrupts bonds of friendship, distancing the vegan from the group and returning them as a stranger.

The multi-layered performances have meanings connected to the revelations of truth that have inspired these social and spatial renegotiations, even when these recalibrations require accepting painful and lonely experiences. This self-surveillance and disturbance are entangled with increasing visibility of consumption that affects personal and professional spaces. Freya, a vegan in her early twenties, discussed her ease and discomfort in veganism as contextual, 'If you're vegan, you surround yourself with like-minded people and you follow stuff on social media and you're always trying out new food. Then when you go out, it isn't a big deal. But when I go out somewhere and it's not like that, I don't know what to do. Recently, my manager at work took out a few people to lunch in a nearby pub and they didn't have anything that was vegan, and everyone else wasn't vegan, so I sat and had an orange juice.' Freya found it easy to transform her personal spaces, from what she sees on social media to the careful selection of places to eat, but this clashes with an impromptu professional outing at work where her food choices are limited.

Speaking to Freya, she was quiet and nervous, 'I'm quite a shy person. I find it hard to make a big deal.' She quietly ordered an orange juice to avoid conflict. The pressures of the workplace and being marked as somehow different, as well as the inevitable questions and surveillance of what she ate (if she ate) were not up for discussion. Yet under different circumstances, she has had confrontations, especially with her best friend who 'doesn't really see the point in veganism, so we don't really talk about it. We argue sometimes which is kind of sad. I feel like it is becoming easier now with stuff like vegan ice cream. … but [I'm] so sick of speaking about it all the time and having to go out and be questioned. I just want to stop speaking about it.' The disturbance of friendship by arguments and confrontations over veganism is frequent and upsetting. When Freya became vegan, her friendship longer held a shared frame of reference and disrupted traditions and activities, distancing what before was commonality into difference (Simmel, 2008). This distancing is felt most acutely in the disruption to rituals of sharing food, but with increasing choice in vegan versions of the same product, this space for confrontation and conversation is potentially minimising the politicisation of friendship.

The visibility and normalisation of vegan consumption expose greater numbers of people to the ordinary lives and practices of veganism, moving them closer in similarity and in proximity to consuming animals. However, the removal of friction removes the centring of the pain of animals when violent eating practices and veganism can exist alongside one another. In 2019, the EU banned vegetarian substitutes in burger or sausage shapes from being named burgers or sausages, with suggestions instead to name them 'veggie discs' or 'tubes' suggested both in comic and serious forums (Stone, The Independent, 2019). 'Burger' does not pertain to specifically one animal's body made into a circular shape inserted in a bun but rather is a cultural phenomenon closely tied to identity politics (Adams, 2018). The linguistic separation of vegan foods reaffirms them as out of the ordinary because they threaten the historical and social distance from who we are consuming. To render veganism invisible through fluid and un-conflicting consumption within spaces and institutions of violence is counterintuitive to its collective and worldly imperatives.

Simultaneously, rejecting places for vegans who consume in spaces of violence – places which also have the power to make vegan options cheap – is classist, preventing accessibility, and can be framed as antithetical to veganism itself. Inevitably, different compromises are made by different vegans in varying navigations of the world they now inhabit, perhaps leading to a consideration that rather than conflict, a far more generous, caring and connected approach to disparate opinions is necessary.

Sheila discussed this with me in the context of her own decisions and choices on navigating non-vegan spaces that make her uncomfortable in order to present an accessible veganism. She said, 'We know vegans who won't go out with family or friends unless they go to a vegan restaurant. If I put those demands on my family, they would say we won't invite you. People would think, 'imagine being like her, she doesn't even come out with us for birthdays because she doesn't want to see someone eating meat.' By going and laughing along, they think 'it can't be that hard, she's here with us and seems happy.' If I was angry and ostracising myself, then I would probably be living up to their stereotype.'

When anger becomes isolation, the removal of oneself from spaces of friction is a loss of potential, ending friendships that it may have been possible to transform. Sitting in and with discomfort is part of veganism. By showing up, your presence disturbs the normalcy of eating animals and makes normal the refusal to do so. Truth cannot be performed in absence and the silence of non-attendance may echo at first but is soon forgotten. The vegan at the table is less frequently ignored, being and feeling watched as a representative for not only animals, but for defying stereotypes, challenging ideas, and engaging in dialogue.

There remains, however, cautiousness in revelations of veganism and the process of engaging in continued conversations in relationships as to why a person has become vegan evokes a fear of how they might be accepted, or not. For Rhys, quoted in Chapter 4, this feeling meant he remained quiet, about telling his family he had become vegan, even though they were vegetarian. Talking about this, he said it was because 'food is such an emotional thing and the ideas of going home and having a meal cooked by your family are bound up in that.' These fears of exclusion influence the choice (not) to disclose their veganism. This avoidance is not sustainable, but rather eases people into navigating their performances around others, knowing that once they do disclose their new practices, they will be watched and feel pressure from external sources.

The fear of rejection disturbs relationships. For Shane, his world and self were disturbed by vegan truths, forcing re-learning of how to negotiate and perform identities felt as central to who he is becoming: 'veganism changed my relationship with myself, which consequently changed my relationship with others.' The realisation and education that veganism is a personally, collective, and worldly good often means attempting a radical break with the past. In the past, there is a self who did not care and inflicted harm and suffering, but vegans in these interviews also referred to a past self who was not engaged in pain and suffering. Their past violent practices were understood as learned through socio-cultural norms and practices. A desire for a return to the pure self of a (childhood) past was expressed in these powerful imaginations not only as something to strive for, but for something to return to, in a temporal manipulation of the future to become also the past.

Recurring truths

Many stories in animal activism are recurrent (such as the encounters with animal death at the opening of this chapter). This is not coincidental but consequential of the conditions of spaces and relations where people care about animals and commit to veganism, the networks of friendship which becoming vegan draws people towards. Somatic and proximal closeness disrupts spaces in spatio-temporal truth encounters beyond the human. This cyclical nature of encounters is part of a larger trajectory of caring for animals, where individuals believe these encounters to be exceptional transformations of everyday life. These mundane revolutions of self are part of a rhetoric of returning to 'the truth' through new forms of embodied knowledge.

This feeling of return is present in vegan narratives in a non-linear temporality whereby a 'true self' is hidden by social and cultural conditioning of violence towards animals. For example, Shane said, 'I've always loved animals since I was a little kid. Before I had a pet or anything, my earliest memories are of the neighbour's kitten, they brought them round the house, earlier than I probably even remember my mom and dad. Sometimes you just lie there and think, were some things just meant to be?' Shane talked about his upbringing in a dairy farming and hunting family in Ireland, which he has always been opposed to, and the fractured relationship he now has with them. This follows his confusion, hurt and anger over their relationships with animals since he was a child. His felt complicity and out-of-placeness with his human family are retold as resistance through his affinities with animals, a feeling justified when he 'found' veganism and undid his 'social conditioning.'

Following Pfeffer (1965), freedom is not in opposition to recurring cycles of time, but rather a wilfulness towards the continuation of worlds. These narratives of recurrence allow for vegans to say: 'this is the way I should always have been' (Matthew). Where space is open-ended but finite and time is infinite, eternal, and immutable, their circulations create echoes. What was, what is, and what will be are entangled and intensifying cyclical continuations of one another. The most powerful example of this recurrence as a spatial metaphor and temporal storying came from Matthew, who initially went vegan for health reasons but soon embraced the political and ethical elements of animal rights and environmental improvements.

'In Newcastle, we have something called the Town Moor which is a huge moorland right within the city centre. The Freemen of the city are allowed to graze cattle there, so there's all these cows in the city centre. The cows are only on the Moor between March and November every year. In a weird sense, I never really thought about it. [I thought] the cows are out here during the nice weather and then they put them inside in the winter. But they're not dairy cows, they're beef cows, and obviously its new cows every twelve months. We walk our Great Dane there. He loves playing with these cows and because he's not a little terrier who chases them, he just walks amongst them and they like him. Anyway, there's this weird moment where the first walk after I turned vegan, when they came back again – huh of course they didn't come back again – a different set of cows came back. It was emotional when we saw them, my wife and I and we just had this kind of moment. These aren't the same cows who we built a relationship with. The cows came out and they were skittish and jumpy. I thought, "that's weird, they were fine with us last year." My wife said, "you know these aren't the same cows?" It was a moment of realisation: oh my god this is horrible. It's so horrible because in a city you don't often encounter the meat that people consume because they live elsewhere, so there's this disconnect. So, we've got this new set of cows and we're building up this bond with them again. I was out there this morning for 45 minutes, and they were lying down, and they don't mind us being there. I've got this newfound bond with these cows, which I think I always had but because I now have this stark new reality that come November, they're going off to slaughter' (Matthew).

Time is spaced by expectations of returning cycles of the stories of the spaces we inhabit; Matthew's cows returned to him year after year, until they didn't. When he meets these cows again, they don't remember him because they have never met him before. Where Matthew previously perceived their behaviour as forgetting and refamiliarisation, after the truth of veganism, the recurring cycle is broken. The human-cow(-dog) relationship is haunted by the coming return of the future, ending in the violent death of the cows. The ethical implications of non-linearity of space and time across an elongated timescale have multifaceted effects within and beyond the human.

The stories we tell ourselves of time and space are related to how and where people care. To search for these echoes is a move towards understanding the universal through the personal. These narratives of return and recurrence are powerful in veganism as they connect spatio-temporally atomised events and gatherings with the social, political, and cultural experiences of disturbed worlds: Matthew's world shifted somehow, and he had to renegotiate across historical and future space and time.

Disturbing truths

Ideas of recurring stories and returning to a 'truer' self is helpful in understanding how vegans are understanding and deploying truth in the narratives of their lives. Pfeffer (1965, 297) describes recurrent temporality as a conception in which 'what already was and will return is this present moment with its power to make the past vital and its possibilities to create the future.' This opens space for thinking about how creative transformations enact different futures, and to understand how vegans narrate their transformation. In contemporary veganism, the body is closely entangled with beliefs about other animals, and this entails holding responsible past and present actions against a future that holds the possibility for others to return to themselves. Activist practices of education and outreach create the conditions for the expansion of veganism.

Shane points towards the ways in which humans are 'so creative in the ways we torture animals. Putting holes in cows to feed them because it saves time crossing fields and stuff ... turning lights on and off in cycles to increase hens' reproduction. Some people don't want to talk about that because they are unable to accept the reality. Imagine never being comfortable and never having anyone there who could sit next to you and hold your hand.' The creative capacity of humans is enacted to maintain the abuse and subservience of those positioned as 'less-than-human.' Reduced only to a body, eating animals disturbs traditional relations between humans and other animals. This confrontation breaches space and time and produces 'the truth' around which vegans organise and act as a singular shared vision when, in fact, this knowledge is contested, shifting, and personal.

Contemporary veganism produces self and community surveillance under the pressure of not doing 'enough' of the 'right' things (Craddock, 2019). Concerned with this ever-present sense of responsibility, the pressure to perform veganism is both external and internal. The ideal vegan is not a real figure but rather a construction assigned to others against which all are bound to failure. This discomfort in aspirations towards an ideal is part of 'an affective encounter between our bodies and the audiences, objects and spaces which we negotiate' (Johnson, 2017). Where veganism has been dictated by the white, male bodies of the powerful, the movement reflects these same exclusions. Even though the movement is open, it is not equally so, and not everyone within it can find the same belonging.

For many, this disbelonging is external to veganism, with a distancing from previous social relationships and spaces, like Sheila who 'went to a predominantly white middle-class school and I was a white middle-class girl. But that shifting of suddenly becoming a minority ... and having to adjust to that that is quite a culture shock.' Sheila's felt difference in society was an adjustment, especially in eating spaces. These social spaces prove particularly threatening to vegans and their strangeness and disruption is heightened here. Some vegans found their exclusions and empathies to transform their perception of other oppressions, envisaging animal, and human violence as resulting from similar violent spatialities and structures (Springer and Le Billon, 2016). For others, veganism itself reproduces raced and gendered societal exclusions.

The connections between animals and feminism (Adams, 2010), farming and the holocaust (Spiegel, 1997), race and gender politics (Harper, 2010), vegan and feminist advocacy (Ko and Ko, 2017) and the institutions of prisons and slaughter as raced and animalised places (Morin, 2018) have been theorised from different positions and contexts in the West. In some areas of veganism, there has been a move to intersectional thinking, but the figureheads of veganism remain largely white, middle-class men, including in the ostensibly more diverse space of social media. In her interview, Esme recounted her experiences as a black woman in the vegan movement and its intersection with feminist and anti-racist work: 'I am friends with loads of people on Facebook who are vegan. If there is a vegan event, the majority of them will click "interested" or "attending." Then I'll see an anti-racism march happening in London as well and there'll be one vegan Facebook friend who is attending. Everything theoretically means that they should be linked, but I think it's really easy to be a vegan and only care about veganism. I wouldn't say I'm particularly active in other social justice movements. I'm very interested in them because it is my life as a black woman, but I don't feel guilty when I'm not doing activism for them.'

Veganism is not a homogenously experienced identity or practice, yet vegans often present the relationships between humans and animals as having a fundamental 'truth.' In this chapter, the circulation around vegan 'truth' has raised questions about the activism, networks, and geographies of veganism that allows us to understand the personal and embodied experience of veganism, but also how this narrative of 'truth' allows for renegotiating temporalities of the world through veganism.

Summarising Part 2

In Part 2 of this book, I have drawn on interviews with vegans based in Britain to explore two interconnected things: first, in Chapter 4, the historical legacies and contingencies of animal activism, the spectrum of tactics, and the proliferation and successes of quiet activism approaches tentatively offered in interviews with vegans. Then, in Chapter 5, I connect variegated stories-so-far of humans and animals within elongated spatio-temporal networks through different temporal feelings of, performances of, recurring and disturbances of truth to explore how embodied knowledges are entangled with the circulation of and around 'truth' in veganism is explored as encompassing variously a sense of injustice, a disturbed sense of place, and distressing (bodily) implications of pain and death in veganism. Both chapters are informed by and inform cultural and social geographies from the scale of the body, and the importance to 'world-making' and movements through space of embodied knowledge. This contemporary veganism is not divorced from the historical geographies of Part 1.

Thinking with veganism works in the mode of 'staying with the human trouble' (Haraway, 2016), where the (human) body is the medium through which we understand plural and entangled pasts, presents and futures. Following Yancy (2005), the body is a contingency, a battlefield, constantly remade, across history and space, both human and animal. Historical and contemporary animal activism seeks to open the beyond as a space for relational connections between here-and-now and there-and-then, in sites of active engagement (Fienup-Riordan, 2000). Moving beyond the individual body to the collective body, veganism might be understood within a multispecies social flesh and the matter and material that constitutes this social flesh, 'built itself from externalizing its expansion into and onto the bodies of others' (Povinelli, 2018, np).

In understanding veganism's connectivity within, between and beyond animals, humans and the world, vegans feel ethically and politically compelled to centre animals and the world outside of anthropocentric imaginations of the future. The embodied manifestation of truth enacted through refusals and openings of relations and networks through their politicisation, which encompass and affect both humans and animals. Veganism relies on a reconstruction of the beyond: beyond archives, beyond animals, and beyond the present. This beyond is the horizon and frontier of pasts and futures already unequally and violently excluding some from the present (Povinelli, 2011). Theorists and theories struggle to describe what exists in this beyond space that decentres the (white, Western, masculine) human and disrupts dominant ways of hierarchical being (Povinelli, 2018) where other modes of existence have been dispossessed. A multispecies social flesh implicates within it those haunting just beyond this chapter even in their absence. Ecofeminist (Mies and Shiva, 2014) and critical indigenous theory and practice (Todd, 2015) decentre the human subject in 'history, governance, cosmologies and legal orders' (ibid., 231) that foreground new modalities of living beyond the human.

Part III

Futures

Figure III.1 Chickens speak. Copyright Catherine Oliver 2017.

6 Chicken-human relations

How to write about and for animals is a central concern in animal studies (DeMello, 2012), both in representing the relations between humans and animals, but also between animals themselves. Throughout this book so far, animals have occupied a space just beyond the boundaries of the writing: the absent referent around whom we are orbiting but have yet to reach. To think about animals differently seeks to rethink who counts in the collective 'us.' Where the concept of 'animality' serves the metaphysical primacy of humans (Cavalieri, 2009, 3), it is necessary to find ways around and through the presumed 'barrier' of species. In my attempts to do this, shared over the next two chapters, I slow, watch, and listen towards and for animals, rethinking our togetherness as a 'quasi-us' (Serres, 1982, 88).

A posthumanist ethic of recognition – 'I recognize you means that I cannot know you in thought or in flesh... I recognize you goes hand in hand with you are irreducible to me, just as I am to you... I will never be you, either in body or in thought' (Irigaray, 1996, 103) – begins in the negative: 'I will not' becomes 'I am not all' becomes 'I am not so the other may be' in a provocative call for the removal of humans from animal worlds, practically and conceptually (MacCormack, 2014). Recognising who we are not is a recognition of sameness and difference existing at once, human and animal both absent and present. An attempt to decentre the human might ask (how) can we become beyond (the) human. These are the questions that define the narrative arc of this part around the beyond-human temporalities of a multispecies world.

When I was researching for this book, I went to Berlin several time as part of field trips for undergraduate students. On my last visit, in 2019, I visited The Jewish Museum with a friend. This museum is designed around axis and voids: the axis of continuity is intersected by the axis of the holocaust, ending in the memory void and the axis of exile leads to the garden of exile. The architect, Daniel Libeskind (1991), calls the design *Between the Lines*, with only the straight line, the axis of continuity, being visible looking at the building above ground. Futures ended are hidden, invisible except for those within them, in the axis of exile and the axis of holocaust, in a design that seeks to represent the unspeakable and the unthinkable in its disorientations.

Thinking about temporality, trauma, and what will not be is vital for politicised projects of memory and history. Temporality endures beyond these moments of rupture as sites of revelation for underlying and ongoing truths in the axis of continuity (Shaw, 2010). In the design of the museum, multiple truths and ways of being are lost on different trajectories of history and the future. Thinking towards these multiple possible futures meeting at the present point of intersection, the refraction of the past becomes clearer. Learning across historical trauma can inform struggles against other oppressions within constellational mutually constituted logics, whilst remaining distinct in its context and consequences. The design of this museum, its spatial logics and dis-continuities stuck with me as I began to think about temporality in my work.

Some futures are already exiled to the 'absolute elsewhere' at individual, collective, and worldly scales. They are unimaginable. This absolute elsewhere (or unimaginable space, the space beyond us) is not necessarily a universal metaphysical or moral frame of reference. Eating a chicken extinguishes their subjective future but it also prevents the establishment of other forms of world shared between eater and eaten. If eating a chicken is the site of an event, this eruption is informed by the axis of stories of space so far. When we refuse to eat a chicken, we renegotiate and open a future where chicken and human are not eater and eaten but can move beyond this binary. The chickens whose axis intersected with my own diverted them (and me) away from spatial and temporal exile and into a shared continuity.

Friendship shaped this chicken-human world, within and beyond the multispecies space itself, which in turn circulated around the 'truths' of my veganism. Despite its conceptual and practical flaws, seeking an imperfect and open friendship remains integral to transformative selves, collectives, and worlds and re-centres a hopeful and engaged spatial ethic. Thinking about the 'beyond' requires both histories and futures kept open in a present that is unfinished but finite (Nancy, 1990). The beyond, as a spatial imaginary, messes with linear temporalities, within 'a long history and ongoing set of violent extractions, abandonments and erasures of other forms of existence' enabling a denial of its own violence (Povinelli, 2018, np).

This chapter is structured around multispecies ethnographies before, during, and after humans and chickens met. The lives we share (and don't share) are carried with me in an open and fluid approach of friendship extended beyond the human, renegotiating 'truth,' and offering new senses of beyond-human temporality.

Why chickens?

There is little natural, evolutionarily speaking, about a chicken laying an egg daily. The Red Jungle Fowl, of which the *Gallus gallus domesticus* is a subspecies (Smith and Daniel, 1975), lay 10–15 eggs a year in clutches of four to six chicks (Capps, 2014). It is only with long-term human selective breeding interventions that hens have transformed into egg production machines. Where domesticated species are supposedly thriving because humans 'have entered into a social contract with these species, based on our supposed mutual advantage; we provide and care for them, and in return they feed our soil and give us their flesh' (Taylor, 2011, 208), this 'success' is one of physical and psychological turmoil for domesticated animals. They live bounded in a liminal space of abjection, held always between life and death. Moments of joy are not necessarily absent from the lives of all animals, but to reach these moments we must turn to the specificities of lives.

In my work in the archives and interviews, I felt that I had become distanced from the lived realities of animals. Coinciding with the end of my archival research in April 2017, six chickens came to live at my parents' home. In the archives, a history of 'not-quites' and 'almosts' had produced overwhelming and disorienting in the loss of relations in the present to 'other bodies, to actions, and to situations' (see Bissell and Gorman-Murray, 2019). But meeting Lacey, Bluebell, Olive, Cleo, Winnie, and Primrose was a stark opposition to the stillness and imaginaries of the archives. Their fleshy, embodied lives demanded attention, rather than reflection.

I spent time with the chickens every few weeks, ranging from a few days to a couple of weeks at a time. Mostly, I watched, fed treats, collected an egg or two, tucked and locked them into their coop at night. Often, I had to run across fields in wellies whilst trying to redirect Bluebell or Olive behind the chicken wire fence they've somehow got over, under, or through, ushering them back to safety away from the circling buzzard or perhaps a waiting fox. Chickens' *umwelt*, as living, being, becoming agents of their own lives, is far removed from mainstream human perceptions of them.

To live with someone is to entangle your life with theirs and their world with yours, to try and understand who they are; to live with and love chickens is to orient towards their lives as valuable, in relation to but still exclusive to your own. Befriending chickens is to approach from a wholly different angle than studying historically. When Godfrey-Smith (2016) met an octopus, to understand the meeting, he had 'to go back to an event of the opposite kind: a departure, a moving apart' (5). This history of encounter, that culminated in my departure and distancing from animals, led to these chickens as a meeting of species, reorienting relations of visibility and recognition (Hovorka, 2012).

Theories of relating must reach for absolute otherness and irreducible difference, where the very existence of the other implies their knowability based on self-similarity of existence: 'selves relate the way thoughts relate: we are all living, growing thoughts' (Kohn, 2013, 89). When moving away from scientific rationalism, 'we begin to lose sight of ends altogether. Disenchantment spreads into the realm of the human ... we come to suspect there are no ends and hence no meaning – anywhere ... ends are not located somewhere outside the world but constantly flourishing in it' (ibid., 90). Rethinking ends as intrinsic to life requires unthinking and expanding us beyond the human. In more-than-human work, ends are left absolutely open. Here, beyond the human, ends are not incompatible with openness, but rather help to orient movement towards life that continues to flourish over and over again. Ends have purposes, especially when they are challenged by continuity of life, rather than the expectation of death. These chickens' lives were meant to end, but they did not. Ends are negotiable, malleable.

Rather than the end of species, an approach of friendship and opening 'truth' renegotiates ends, by requiring an attentiveness to the unique worldly beings and knowledges of animals. Specifically, this requires a smaller field and a closer proximity with animals to understand them on their own terms: 'The world beyond the human is not a meaningless one made meaningful by humans ... rather, meanings emerge in a world of living thoughts beyond the human' (Kohn, 2013, 72). Relationships with companion animals, in their closeness to humans, are 'clear candidates' for not only friendship, but 'full, even supreme, manifestations of everything friendship needs to be' (Townley, 2010, 45). Such a feeling does not extend to the animals we eat. For Jordan (2001), the potential of animals as friends should ethically prevent us eating them, but he is less concerned with the specificities or spatialities of these friendships themselves.

Galline histories

The domesticated chicken, *Gallus gallus domesticus,* dates back thousands of years, with the time of domestication assumed to be around 3000BC, in India, Burma, or the Malay Peninsula (Smith and Daniel, 1975). Chickens, both hens and cocks, have occupied multiple positions in many historical and contemporary societies not as the metabolic labourers on the margins that they are today (Barua et al., 2020) but as valued members of multispecies communities. In *The Chicken Book*, Smith and Daniel follow the chicken across space and time to explore how they moved from revered and agentic beings to a situation where the commonplace bird has become, 'under the auspices of our technological society, and one must insist on this, *they will not be chickens and their eggs will not be eggs* (299, emphasis in original). Almost fifty years after their writing, their insights are easily applied to the contemporary ongoing biotechnological manipulation of chickens' bodies in the food and medical sectors that is common, if hidden, knowledge.

In July 2020, KFC 'admitted' that more than one in three of the birds they raise for food in the United Kingdom suffer from footpad dermatitis, a painful inflammation caused by poor ventilation and hygiene standards, which prevents birds from walking normally (Levitt in The Guardian, 2020). One in ten suffer with hock burn, caused by ammonia, which burns through the skin of the leg. The moral outrage at chickens being unable to walk is also couched in the knowledge that 'nearly all the chickens reared for KFC are fast-growing breeds that take just 30 days to reach slaughter weight.' The passivity of language in such reports obfuscates the deliberate biotechnological interventions into hens' growth, or their production for ex-layers. The biotechnological manipulation of laying hens is documented in Karen Davis' *Prisoned Chickens, Poisoned Eggs* (2009, 44), where they 'are not bred for their flesh, so when their economic utility is over, the still young birds are disposed of as cheaply as possible.'

Smith and Daniel's assertion that *these are not chickens* is misplaced, because these undoubtedly *are* chickens, but they are chickens who have been shaped by histories of industrialisation and capitalism working and remaking their bodies. However, biotechnological manipulation of chickens has been undertaken before industrialisation. In 4th Century B.C. Egypt, we find 'a mass society which mastered the technology of large-scale incubation ... four thousand years ago the Egyptians invented incubators capable of hatching as many as ten thousand chicks at a time' (Smith and Daniel, 1975, 14). There is a historical relationship between urbanisation, exploitation, labour, technology, and chickens that sought to remove the chicken from the reproductive process in order that she might be freed up to produce eggs, rather than hatching her chicks; 'since the productive period of egg laying for a hen is not much more than two years, to allow a hen to set and hatch her own chicks would be to use up a substantial portion of her productive life' (ibid., 238).

Closer to home, temporally speaking, Landecker's (2013) insights on the metabolic disruption of night and day have long been used in industrial egg production to conform and elongate chickens' 'productive hours.' The discovery of the direct relationship of egg production and lightness where lightness stimulates the hen's pituitary gland in turn signalling to the ovaries in the years of the Great Depression in the United States was fatal to the hen, this metabolic disruption increasing their egg production whilst weakening them, making the susceptible to disease and, inevitably, confining them to the factories of industrial agriculture.

Chickens cannot escape their inherent *chickenness* even when reduced to their re-productive labour and to their bodily processes. But, as those who know chickens can attest to, when offered the right care and space, ex-laying chickens are capable of learning and finding more *natural* practices: 'the birds nest in soft piles of hay, dried leaves, or curled wood shavings... At Chicken Run Rescue, they dig in the dirt, perch on branches and sprawl in the grass to soak up the sun through their wings' (Hepperman, 2012, on

Mary Britton Clouse's human-chicken household, 23). In fact, those behaviours and diseases that are controlled or punished for industrial hens, such as pecking or deformities, are directly caused by chickens' attempts to find space to be themselves; chickens resist and remain '"wild" in the face of machines that seek to make them docile' (Wadiwel, 2018, 528).

For Beldo (2017, 108), framing the lives of chickens as *labourers* 'allows for the possibility of agency on the part of farmed animals that includes more than just resistance, disruption, or death.' The temporalities, spatialities, and histories of humans and chickens through industrialisation are intimately entangled and, as such, we are required not only to refuse to participate in beyond-human violence, but also to imagine and cultivate alternative spaces of co-living alongside animal activism (Donaldson and Kymlicka, 2015).

Meeting chickens

When researching animals, there is often a 'troubling expectation to bear witness to violence against animals and do nothing' (Gillespie, 2019, 3) for activists and researchers. This ethnography refuses this depoliticisation and began with doing something: rescuing six chickens from a battery farm. I contextualise the chicken-human bond within a landscape of other species, domesticated and not, including cats, a dog, horses, and a pigeon, as well as buzzards, toads, mice, squirrels, and rabbits. Unthinking the human (Dinker and Pedersen, 2016) is an act of de-centring necessary to re-learning multispecies worlds (Hamilton and Taylor, 2017).

The chickens lived in a coop at the back of my parents' house (Figure 6.1), with a wire run attached, built by my brothers a couple of weeks before they arrived. The chicken coop was unused before this, leftover from the previous owners. In the shed next to the coop, there is the previous owner's framed copy of a poem dedicated to shooting game birds. Recalling the sweet pheasant who passes by, I shudder and throw it away, pushing the violent history of this place away. Tranquillity shrouds violence. The run opens – if someone is around to shoo away buzzards – onto a gravel yard. At one end, there is a small lawn, a linden tree planted in the middle, and small fruit bushes around the side. This space backs onto a field where horses graze and the coop leans onto a garage.

In the garage, beds, food bowls, and toys are stacked on a shelf for 'feral' cats to sleep in, the door left open a few inches. Across a path there is the house where Percy, a racing pigeon who found his way here when he lost or left the flight he was supposed to be on, can usually be found perched on the roof. In the house Charlie (a dog) sleeps through the day and kills his toys in the evening, and Fizz, a 21-year-old house cat snores loudly in a patch of sun. Fizz, mostly deaf, became a house cat at 17 after many years of adventures and seems mostly happy, becoming distressed on the rare occasions she has found her way outside through an accidentally open window.

Figure 6.1 Chicken coop. Copyright Catherine Oliver 2017.

Chickens didn't ever much cross my mind before I became vegan. Not exactly the most charismatic creatures (Rothfels, 2002, xi), chickens were both distant and distanced, absent from my daily life in the city I lived in, and the countryside I grew up in. Their archaic dinosaurian feet were more monstrous than endearing; their flightless wings strange and distorted, rather than soft feathers bathed in dust; and their red combs too fleshy, rather than revealing feeling. I would never have even known these were my previous feelings towards chickens, as they didn't cross my mind, share my world, or exist outside of abstraction (although, of course, they did). Only after learning about veganism did I think about chickens. Before veganism, thinking, or trying not to think, of chickens as the layers of eggs for my breakfast was the extent of my attention. Knowing that an egg came from a chicken is worlds away from knowing the processes of pain, places, and bodies implicated in this egg. I purposely pushed away a truth just beyond the horizon of my knowledge, until I couldn't (McDonald, 2000).

An egg, its shell coloured in the middle of pale brown and white, contains a yolk coloured somewhere between a pale yellow and a deep orange and an albumen (white). The chalazae (ropes of egg white) hold the yolk centrally within this shell. Eggs have cultural significance in historical and contemporary religions, traditions, and ritual practices, as well as being symbolically important in celebrations as a symbol of life, rebirth, and rejuvenation containing nascent life in its rounded shape without ends nor beginnings. Laying hens are defined by their eggs. Their bodies and lives are valuable only in terms of the quality, quantity, and speed at which they can produce. Beginning with the egg is not to re-affirm this, but to contextualise what is being resisted. This unexceptional egg takes on a different meaning and matter here, entangled with life and death.

Becoming vegan transformed my feelings about and towards eggs, revealed as products of pain, death, and exploitation. The egg is a holder of violence: unending, cyclical violence. In the United Kingdom, chicken farmers 'must make sure that your flock has continuous daytime access to open runs (mostly covered with vegetation). Runs should have a maximum stocking density of 2,500 birds per hectare' (DEFRA, 2019). Exceptional circumstances, such as bad weather or perceived risks to bird health, can mean chickens are kept inside indefinitely. No longer the symbol of life and rebirth, eggs are the holders of death. When we consume eggs, we consume their violence (Mol, 2008). With this knowledge, eggs become sinister and unworldly: opposed to a re-imagining of a future in which humans and chickens may (be)come together.

When I see eggs in the supermarket, my stomach churns. The towers of cardboard egg holders filled with those delicate shells, holding sticky liquid, disgusts me. Eggs as (human) food have become so distorted and violent in the processes and deaths implicated in their consumption, that I have a visceral reaction, even more pronounced when I encounter the smell of them boiled or fried or scrambled. There is something so alien about them. Something out of place in this sanitised space of the supermarket. Reacting to eggs, I feel disgust and anger in my body towards the places and processes that make eggs food. It is not the eggs themselves creating this reaction, but their alienation from where they should be. An egg holds meaning, to me and beyond me. Before I met chickens, I went from eating eggs to disgusted by them, enacting particular relations to chickens defined by their eggs: consuming them and then resisting them.

When Alice Walker chose to live with chickens in her later life, she asked, 'who knew what would happen next? Who could guess? That I would fall headlong into a mystery. That I would find myself pulled into the parallel universe all other animals exist in' (2013, 5). I similarly fell into a mysterious chicken world. Before I entered this new world, I thought it might solidify my veganism and that through getting to know chickens, I would become a better or more authentic vegan. Acquiring lived experience with food animals was an intended improvement of myself, my activism, and my work, exchanged for housing, food, and protection from predators. My initial approaches to these chickens instrumentalised them in the work I wanted to do, the things I wanted to say, and the futures I wanted to imagine. Once I got to know them, they spoke to and with me in the ways that animals we know well do: through their noises, actions, movements, and behaviours.

Meeting chickens challenges complicity in harmful systems and entanglements with animals and it matters that they were chickens. 900 million chickens are slaughtered for food in the United Kingdom annually and 65 billion globally each year. While there are particular situated knowledges and practices of living with chickens (notably Hovorka, 2012), this close (be)coming together with food animals is unusual in Western, especially urban, societies. Chickens are not present alive in most of our spaces. The naturalised and naturalising representation of chickens' subservience to the human is affirmed by the dissociation of their flesh with life (only with death). It also matters that they were *these* chickens. Their personalities, relationships, and temperaments were unique, and allowed me to get to know them as individuals, rather than as symbols or holders of a violent galline history. Thinking about this animal rather than the animal is a movement towards moving beyond the binaries proposed in Chapter 1 of this book.

Being with these chickens forced confrontations with species-specific oppressions. Species is a blurred spectrum of different kinds of bodies, in a human ordering of other beings that allows us to speak about different kinds of animals and their lives. This allows for the privileging of humans over animals but incorporates hierarchies of some kinds of animals over others, and some individual animals over others of the same species. This is determined by their closeness to particular (powerful) humans and has geographically determined gendered and racial politics. In this construction, species is not a natural designator of value, but rather a socio-cultural predisposition to particular animals through our relations and the contexts of living together (or not) (Joy, 2009).

Chickens are maintained as worthless even as they are integral to multispecies worlds. When I met chickens, I had been attempting to unlearn violent knowledges about chickens through veganism. I still had much to learn and unlearn, worlds to unmake and remake. So, what happens when chickens, their lives, and their spaces are prioritised? Meeting Lacey, Bluebell, Olive, Winnie, Primrose, and Cleo was the beginning of a different kind of fieldwork and different kind of living in the world. Angry as I had been in my work, activism and with the world (Oliver, 2020) and holding onto the importance of that anger, meeting them allowed me to redirect that anger into more affirmative situations. Their lives and deaths in the spaces we shared are a practice in what a different world and future might look like.

As strangers (Ahmed, 2000), chickens rupture social life and disturb place into a space of self-reflection and political engagement, seeking to embrace the imaginative geography and geography of imagination of friendship between individuals and species. Moving towards more hopeful understandings of resistance, the human-chicken relationship is precarious but full of potential. Following more-than-human entanglement experiments (Gruen, 2015; Tsing, 2015; Giraud, 2019) and interspecies relationality work (Haraway, 2008; Willett, 2014), something somehow togetherly can instead be attempted, weaving past, present and future together across individuals, species, and networks. This small multispecies world building can help to understand the futures we are leaving behind. The future as such signifies the end of what was, and orients towards what and who we are becoming.

Beyond-human futures

When a light flash event occurs in space, its path, potentiality, and possibility are at the moment of occurrence spatio-temporally defined. This phenomenon is, in physics, represented by a light cone in special and general relativity: a spatio-temporal elongation to past and future from each observed present. An event's ability to affect another is determined by whether light can reach the space of the second event before it has occurred; if it reaches after the second event's occurrence, it may occupy the former location, but not affect it. Events at the same time but different locations can't affect one another. The light cone holds the same frame of reference for all observers. Outside of the light cone is the absolute elsewhere: the unknowable, unthinkable, unspeakable, unimaginable. Light cones, as spatial structures, contain all that is, was and can be (observable) and exclude all else as impossibilities of the elsewhere. Light cones might aid in conceptualising the kinds of non-linearity, disruption, and potential futures after trauma that are still possible, outside of exile and the absolute elsewhere.

The future is what we leave behind. We define the contours of possibility. We don't inherit only the physical, biological, and economic, but also the socio-cultural, political, and politicised lives and circumstances. Inheritance and disinheritance are part of historical global uncertainties, precarities, and violence. It is impossible to break with the past, but there can be a necessary snap (Ahmed, 2017). (Dis)inheritances are reflected and embodied in how we live our lives, what worlds we construct, and what we refuse. The futures we imagine and enact must extinguish other futures, through 'non-reproductive labour, as the work you have to do in order not to reproduce an inheritance' (Ahmed, 2019, np).

If the present is the space between the worlds we inherited and the ones we leave behind, then we must inhabit this between space intentionally, seeking to affect change beyond it. Being present in this present is a dwelling in the past and future; caring beyond this present across species and beyond the human in an intentional extension of self to possibility. Inheritance as non-reproductive is entangled with a friendship that continues to weave within and outwith the relational subjects and collectives of this research. Such an approach to the future is concerned with political stances on what, who, and where we might be leaving behind and how we might be able to transform those possibilities. Inheritance and friendship are constituted with, against, and beyond the truth of me, others, and the world.

Chickens are awkward creatures, especially chickens like these, bred to grow quickly for daily egg production. I pass the barns they were raised in when I take Charlie, the dog, for a walk. I've seen them full and I've seen them empty. I don't know what the cycle of chicks being raised is and I'm not sure that I want to. There are ten barns, maybe 100 metres long, maybe 200. I can't really bear to look and anyway, I should be looking at the road. Each barn must have thousands of chicks in there for the first six months of their lives. Then these thousands of hens will be moved on to their next cage, their next life, their next death. Every day must be a death in this place: potentiality and possibility destroyed in the denial of the beyond, in the destruction of the future. The grief is overwhelming every time I think about the suffering now, then, here, and there.

The grief does not reach the space we share together, usually, because being with chickens requires me to be in the present, to attend to this space that we share now in a 'subjective experience' (Spannring, 2019). When I am required to be present and attentive, our histories dissipate, unwelcome in this required 'sanctuary' space, that must be created alongside any animal activism (Donaldson and Kymlicka, 2015). As those who care for and about animals, we must not only transform the present but also 'imagine the contours of just relations that human might have with "farmed animals" once we stop confining and killing them for food' (ibid., 50). This is an attempt at an alternative world, informed by my ethical political beliefs and knowledges. Different to a sanctuary, but offering safety nonetheless, this attempt to remake their unmade world of pain (Scarry, 1985) requires a closeness specific to our space, where there is something uniquely between them and I that is not singularly common (Simmel, 2008).

7 Multispecies futures

I sit on the steps ten feet from where Winnie, Primrose, Cleo, and Olive are pecking on the lawn, searching for worms in a garden planted with linden and fruit bushes for them. They move slowly around the perimeter, especially focused under the bushes where the soil is damper. In the distance, I hear birdsong and cars. Their bodies brush up against one another, an assortment of copper, gold, and white feathers, until they stop and one of them settles into a shallow hole in between two bushes, only her head visible above the grass. The two others are by her side, peering into the hole at the soil she is pecking, and I can see Cleo distracted on the other side of the garden, exploring something near the fence. I wonder if I go towards them, will they notice me? I've been sat here ten minutes without catching their interest, so slowly I walk over and sit down in the damp grass. Primrose stops and turns towards me. From here, a little closer, I can see the yellow ring on her ankle that represents her name. She's disrupted by me coming close. She stares for a few moments before returning her attention to the ground. A minute or so later, she gives up on the patch of soil she is pecking with the others and slowly moves off back to exploring the muddy perimeter.

I watch Primrose move away as Winnie and Olive continue to peck in the shallow hole. Over in the other corner, Cleo is head deep in the chicken wire, not showing it if she has noticed me. Olive is leaving now, fed up with Winnie taking up the best perch, or maybe she has seen something I have not. Winnie's quickening head movements signal she has found something good. I turn away so as not to distract her. I know that they know I am here, but I don't ask for or receive their attention. Do they mind? Would they rather I left?

Sitting here, I can't help but think about Bluebell (Figure 7.1). She loved to escape, sometimes taking a friend with her, but mostly darting off across fields on her own. Since Lacey and Bluebell's deaths, their absence hangs in the air; their soft, quiet deaths at odds with the violence of their early lives. I remember Bluebell's soft feathers when she moulted, and the thick white layers. How when a feather fell, it looked unlike any other. From a distance, I can't always tell who I am watching. Some worlds are

Figure 7.1 Bluebell. Copyright Catherine Oliver 2017.

not for me, but most worlds aren't for them. In this cultivated space of human-chicken togetherness, the silence between us is not empty, but filled with possibility. In this silence, Bluebell returns to me. How can we build spaces together that protect and enable freedom? She always wanted to know what was on the other side of the fence, but out of our worry for her safety, her world was bounded to the finitude of safety. Bluebell wanted to explore the world and she needed a place which was for chickens alone, for her alone.

Negotiating the protection and freedom of chickens requires me to know them as individuals, to think what might be best with and apart from our human worlds. Where unanswerable questions bound our worlds, for Freeden (2015), these questions are silences that contain and transform communication, understanding, and feeling at the absolute limits of the human. There are, then, four kinds of silence: the unthinkable, the unspeakable, the unknowable, and the unconceptualizable, each defined by the subject's location within the limits of collective humanity. The last, the unconceptualizable, refers to the absolute elsewhere.

Our relationship requires me to actively inhabit this space, paying attention to what is and what could be. When evening comes, the chickens have usually already gone into their coop, up the stairs, and nestled into one another. Their purrs are soft now, eyes lit iridescent from the light on my phone. I close the door, bolt the lock, and gently kick the brick in front of the door so no rats can slither in. As I walk away, I think about all their living unknown to me, unapologetically chicken. I wonder if perhaps our desires for the future might not be so different: peace, freedom, friendship. Our socio-spatial fluency and familiarity with animals has been destroyed, making strange and dangerous these animals who inhabit the world differently. Once distanced and different, human treatment of other animals becomes more violent, more destructive, and excusable. Bluebell taught me something about chickens in her determined attempts to escape; she deserved the world she desired, and we must find ways to create it.

Human-chicken friendships

A few months before the opening story of this chapter, as dusk fell, I headed out to check on the chickens, remove any food that might attract rats and to secure their coop for the night. I count five chickens, but one is always missing, and it is always Bluebell. Where this time? Under the ferns, or down by the ditch? Climbing the fence, I trudged through thick mud, glancing ginger feathers warm against the brown and grey of winter. Bluebell's head turns slightly, and I know that she has spotted me coming. Ruffling her feathers, she gets ready to escape me. Are there more worms over here? And doesn't she know that the fox or the buzzard could arrive anytime? I'm a few steps away now and… it's time to run! Cornered, I reach down and just miss her, and she's off through my legs and into the field. As I chase her back and forth over to the coop, I laugh. Occasionally, Bluebell stops to peck at a worm again for a few seconds before I catch her up. Opening the gate, she rushes through, joins the other chickens, and looks back once more with victory in her eyes.

In *Chickens' Lib*, Clare Druce (2013, 218) documents how 'the catching and pre-slaughter treatment of poultry is wide open to abuse' through the example of 'bagpiping [which] is a slaughterhouse "game" where employees squeeze live chickens until faeces squirt out, to be aimed at fellow workers.' The dangerous and disturbing work of chicken catching is increasingly being undertaken by machines, which means chickens no longer meet humans but machines in the slaughterhouse. This reveals 'the reality that chickens are "wild" in the face of machines that seek to make them docile' (Wadiwel, 2018, 528). Machine catching is reserved for the slaughterhouse, and 'intensely frictional' catching by human hands of 'chickens who would prefer not to be caught' (ibid. 543) remains the norm prior to transport to slaughter. The dimming of lights and working in collectives at night to corner chickens are tactics to quash the resistance of chickens to the harvesting machine.

'Every night in the United States, approximately 8,000 chicken catchers put on throwaway suits, rubber gloves, and dust masks. In a few hours the masks will be soaking wet and black with dust, and the men will tear them off in order to breathe' (Davis, 2009, 133). The lights flick off, and the chicken catchers move in, disrupting the silence into screeching, roaring chaos (Clark, in Davis, 2009). The chicken catcher's job is to corner chickens and grab four or five birds by one leg in each hand; the birds are then shoved into a drawer of a cage on a forklift truck, forcing any protruding body parts. Flapping or struggling earns a bash over the head and, although broiler birds tend to avoid the worst of the bruising as it would damage their price, 'spent' laying hens have no value in how they look.

Chicken chasing and chicken catching has a violent place in the space of the human-chicken history. The resistance of chickens to their catching is not only a means of survival, but a means of asserting their wildness, their beyond-humanness, that cannot be flattened and made docile despite centuries of human manipulation and intervention. Every night, as Bluebell and the other chickens here are encouraged into their coop, agro-industrial chicken catchers are putting on their suits, readying to grab chickens and stuff them into cages to be transported to their slaughter. The industrialization of chickens relies on a deep human knowledge of their behaviours and biologies that has been exploited over centuries of living together (Smith and Daniel, 1975). How, then, as I chase Bluebell home, can we work to undo and rethink multispecies futures through intense, local, and personal relationships?

Smith and Daniel (1975, 333) write on keeping chickens that 'as in all good things there are problems and complications. A free life is, for chickens as well as for people, in some ways more arduous and more dangerous than a confined life.' However, almost 50 years later, the relative protections of a confined life for chickens are negligible and the benefits of confinement mainly for economic gains. The potentials of a freer life are yet to be realised and this assemblage of people, chickens, and other animals offers one perspective into what a multispecies future might look like through a situated spatio-temporal friendship. Befriending a chicken, unlike befriending a dog or cat, disrupts their position in normative species orderings and situates practices differently, in light of the elsewhere worlds of chickens and humans.

A friendship is not (usually) fleeting because it requires a sense or feeling beyond a shared location: a holding (of) space. Not a relation of convenience but of chosen ethico-political intent, friendship's boundaries are shifting and contingent. Friendship is open: not necessarily bound to a cause nor a shared purpose, the relation implies something somehow more enduring and its openness means that it can always be returned to. Friendship is an endlessly plural relation that is singularly common; but it is also dangerous, its openness being open to powerful actors. Friendship must resist its own co-option by and for the politically and ethically dubious. A 'commitment to truth telling lays the groundwork for the openness and honesty that is the heartbeat of love' (hooks, 2000, 92), which holds across discussions in the previous chapters. Beyond me, beyond chicken, beyond the beyond we share together – never quite breached – friendship might offer an ethics and politics that make a difference.

Interspecies friendship requires moving beyond entanglement for the semi-permeable boundaries of bodies (inter)facing other semi-permeable bodies in transcorporeality within a porous material world (Alaimo, 2012). When I approach in friendship, I have already declared something in the relation that defines and bounds us. 'Naming is, in a way, the very first and most basic act of language, because it is what enables us to talk or write' (Borkfelt, 2011), and this act carries values, ideas, and conceptualisations of its object or subject. Naming this relation friendship is a political act, to divorce from other forms of relating. Friendship is in the direction of the future, much like activism is constructed a transformative present reaching beyond itself.

Despite the uncertainty of knowing Bluebell's experience, my protection is friendship. Her teasing glance back is hers. Rather than centring relations around only an imagined or actual ethico-political cause as is traditionally the mandate of masculinised comradeship, friendship puts other personal, collective, and worldly demands on the relations, entanglements, and networks of friends. Ever expansive, and always open, friendship includes and excludes, unwittingly and deliberately. The accountability and boundaries of friendship must be redefined and repurposed for less violent multispecies futures.

A shallow pool was repurposed for the chickens' dust bath, regularly replenished with wood ash and dirt, the introduction to a designated dust bathing space was one of great excitement (Figure 7.2). Despite having never met a parent or older hen to show these chickens how to clean and bathe in dust, they knew almost immediately what this space was for and, ready to try it out for the first time, they all rushed to pile in before realising it would be better to take this in turn. Ground-nesting birds dustbathe to clean themselves, remove oil from their glands, and to cool themselves in hot weather (Davis, 2009, 72) and takes about half an hour. All hens, even those in battery cages, try to perform this behaviour on wire floors, their motivation

Figure 7.2 Inside the coop. Copyright Catherine Oliver 2017.

increases the longer they are deprived of their bathing, causing pain and damage troubling the animal (Vestergaard, 1987).

Watching the chickens in their dust bath, my mind conjures an image of their fleshiness beneath their feathers, the body whose death is fetishised for cheap human food; 'no longer are you seeing normal products of everyday existence. In front of you is the violent reality of animal flesh on display: the bones, fat, muscles and tissue of beings who were once alive' (Stanescu, 2012, 568). Chicken advocate Mary Britton Clouse (2015) writes that 'taking a dust bath is the closest thing to heaven for a chicken. They bathe in the sun and in loose dry soil depressions in the dirt which cleans their feathers and helps rid them of parasites.' As the chickens stretch their wings, it is hard to disagree with her.

Humans, animals, and other species already live in a multispecies commu-nity, one that is most visible where we live alongside one another. Chickens, cats, pigeon, and horses are separated partially by fences and wires, but always open to one another. Chickens chase cats who are wary of chickens. Chickens fluffing their wings if the cats come sniffing. One horse is scared of chickens (and dogs and cats) and she will kick out for them if they get too close to her feet. Percy, an ex-racing pigeon, is rejected by other pigeons, who know he somehow doesn't belong. Percy remains here, keeps trying to perch next to them. Domino and Hunter, feral cats, curl up in their beds, some-times joined by Humphrey, but flighty at the sight of the other pair, Ghost and Lady. Watching these animals navigate around and with one another, the articulation of interspecies communication and community alienated from humans appears more intact between other species in our absence.

Watching and writing about the chickens, I am forced to reflect on my meaning to them. I approach in friendship, but am I still a stranger? For the animals I sometimes live with, I appear and disappear. I don't know what this time feels like for a dog or a chicken or a cat. Do they wonder where I go? The relationships between individual animals (me being one of those individual animals) influence how we respond to each other when we come into proximity. These responses and relationships emerge from our histo-ries together, across time, space, and species. Perhaps it is confusing when I intrude for a few nights or a couple of weeks. Perhaps they remember me, perhaps they remember of me, perhaps my return disrupts as the stranger approaches. Abstracted truths become challenged when living in multispe-cies worlds and practising different futures and uncertainty is rejected in the realities of being together (Wadiwel, 2016).

Beyond-human geographies

Thinking and being with those whose futures (beyonds) are returned, through rescue, opens a space to assert the vitality of the beyond (that is a future) in keeping open a finite history rather than a finished history (Nancy, 1990): 'beyond [au-delà]? Which beyond? How could there be a beyond poli-tics? How could there a beyond in any matter whatsoever? And the step not beyond [le pas au-delà]? We are quite familiar with these questions. That beyond, it is necessary to hear it differently – beyond the beyond I could say' (Nancy, 2014, 2).

Beyond-human geographies ask questions of the future through trans-formative visions and relations in the present. Engaging critically with other ways of living, be that veganism or with other animals, requires us to locate ourselves beyond human, beyond animal, beyond this space we share together. Where both human and animal are constituted in their relations, our shared spaces are made meaningful in a politics that 'is as much about "being" as about "doing"' (Smith, 1999, 131). Ambiguities and uncertain-ties, or a politics of doubt (Wadiwel, 2016), maintain the openness of these spaces where the difference that difference makes has already shaped our ethics and politics.

For Braidotti (2018), the posthumanities have a tendency to mourn or celebrate the emergent category of the 'endangered human' creating a simultaneous over-exposure and disappearance of the human which is difficult to disrupt. 'Approaching the present produces a multi-faceted effect ... the sharp awareness of what we are ceasing to be (the end of the actual) and the perception of what we are becoming (the actualization of the virtual)' (ibid., 7), spelling the rhythm of continual subject formation. Even in the absence of the human, the human world will sustain. 'Bringing the animals back in' (Wolch and Emel, 1995) fragments those who care about animals into different kinds, modes, and ethics of caring. The non-human, the posthuman, and the more-than-human are all looking for ways of being and becoming (together) beyond the self and encountering and working with the world as it is and as it may be, refusing humans as uniquely unique (Midgley, 1983).

The day after Lacey died, I wrote in my field notes, 'Something is different today. Lacey died last night. She had been ill for a while, not eating, slow and thin. Tucked up with her friends in their coop, she fell asleep amongst their warm bodies and didn't wake up. I wonder when they knew, how they cared for her, if they're ok without her. They stayed inside today, mourning Lacey. So am I. Our pain is right now beyond shareability, but perhaps it can affirm our futures together. From such a violent beginning, her end was peaceful, quiet and surrounded by friends. This end matters. It is not the end.'

For laying chickens like Lacey, their death would usually take place when she is considered 'spent' by the industry, which is not the end of their laying life but rather the time at which their economic utility as daily layers drops off, at around 72 weeks old (CIWF, 2012). Because laying chickens are not bred for flesh, their deaths need to be cheap and massified: 'some are suffocated to death in dumpsters ... others are gassed and buried dead or alive in landfills or ground up in woodchippers. Still others are trucked to slaughterhouses and turned into meat products such as "spent hen meal," which is fed back to the hens' (Davis, 2009, 44). Rather than killing, these deaths of spent hens are referred to in scientific literature as 'depopulation' (Kristensen et al., 2001), 'management' (Newberry et al., 1999), and 'processing and utilization' (Kondaiah and Panda, 1992). The aim of technological advancements in 'depopulation' is to speed up the killing of chickens and reduce their contact with humans when they arrive at slaughterhouses thus, in their theory, increasing animal welfare (Kristensen et al., 2001). It is in the hands of human workers, paid for speed, that physical pain is exerted on chickens before their deaths by machine. Their bodies are eventually ground up and used as by-products, their 'skin, gizzard, heart, ova, yolk and fat can be utilized in emulsion-based chicken and other meat products' (Kondaiah and Panda, 1992, 255).

Living with chickens creates continual confrontations with human violence towards them. Lacey's, and later Bluebell, Olive, Winnie, Cleo and
Primrose's deaths, were far removed, physically, from the deaths at the
slaughterhouse. However, they could never be entirely disconnected. In
mourning the chickens who I lived with, the realities of the chickens I don't
know is haunting. All that is possible is to reach beyond these present temporalities for a politics that makes a difference, refashioning interspecies
affiliations (Vermeulen, 2017). Here, veganism is not the end point of an
interspecies friendship, but one approach that allows us into the worlds of
other animals and to theorise and cultivate spaces beyond the human.

Beyond everything

In my relationships and research with these six birds, I have been concerned
with the 'beyond-human': beyond them, beyond me, beyond us, beyond
this beyond we share. The beyond is something impenetrable between us,
holding us necessarily apart and together. This beyond space is not only a
holding space, but rather a site for critical inquiry and intervention between
and within multispecies worlds, differently sensing and navigating temporal,
spatial, and bodily distances of continuity and other species' experiences
of them. This beyond is spatio-temporally and socially elongated, allowing
for theorisations of the space between us as one which we might cultivate
to allow for spontaneous encounters and longer-term commitments to one
another.

How do I know what the chickens do when I'm not there? I imagine
it isn't dissimilar to what they do when I am there. My presence at first
was unwelcome – I really wanted them to love me and to care that I was
there. Trying to build a relationship without knowing chickens, I sometimes tried to stroke their soft feathers, or hugged them too closely as
I picked them up from one of their adventures to take them to safety
away from predators. Then, I was welcomed and, I think, recognised as
a friend, or at the very least '*holder of corn.*' I became of interest as I sat
on the grass or steps and let them peck my feet, my hands, my clothes.
I think that this is a space that I belonged. The dimensions, shapes, and
requirements of shared human-animal spaces require ethical and political navigations and compromises even in veganism. To leave animals
alone (MacCormack, 2014) is not a call for a radical break of humanity
from multispecies worlds but a renegotiation of distance, intervention,
and (be)coming together.

A future-oriented veganism requires some element beyond death to sustain it. Friendship requires something similarly beyond the individual. If
veganism holds that people should, as far as is practicable and possible, end
the use, exploitation of, and cruelty to animals, then we must centre this
truth within the specific conditions of our lives, of the world, in the present.

Not relying only on history to guide us, the material circumstances of spatio-temporal encounters in the contemporary world are crucial to imagining how the future might be beginning today, yesterday, tomorrow. Living with chickens has demystified them just enough for the political, ethical, and bodily realities of their lives to be revealed. The myths and stories we are told about animals shrink and expand their representation in human worlds, positioning specific animals, and species as absolutely helpless or absolutely monstrous: as subject or object, being or thing. Veganism does not do away with this binary, as unlearning and unknowing is an iterative process that requires creation of new knowledges experientially, not only imaginarily.

What does the world look like for someone else? What if that someone else is not a human, but a chicken? To share a bed of shavings with five others, bundled close together? To lay an egg every day, or thereabouts, then to have the egg taken away to an unknown elsewhere, never to be seen again? When I first started living with chickens, a friend gave me a book, '*The Hen who Dreamed She Could Fly*' (Hwang, 2013). Sprout, the hen, decides to refuse to lay eggs that continue to disappear to she doesn't know where and plans to escape, hatch an egg, and raise her child. Sprout watches the ducks outside who fly and raise their young: the freedoms Sprout desires. But her coop was built on a slant, so that every day her egg rolled past a barrier out of reach. How far can we imagine the lives of chickens? When I bend over in the yard one day, a chicken found her way onto my back. Bent to the ground, I was stuck at this angle until she decides to leave. I glanced around. Rarely is my head a foot off the floor, but I can see, now that I am closer, the remnants of corn, lettuce, and mealworms invisible to humans. The gravel doesn't look as uneven either. And, is that a hole in the chicken wire that I – I mean they – could escape through?

The world looked different from this direction, head to the ground, the space became one of confusion of a familiar and strange place of disorientation. Veganism requires and follows 'seeing ourselves and the world the way it is rather than the way we want it to be' (hooks, 2000, 68) and making a move to transform it. Not only are we talking about a biopolitics and a necropolitics, but how the worlds of those whose bodies are both (and neither) alive nor dead might be different if approached from another species' perspective. The space of the beyond, of possibility, bounds, and unbounds our becoming-together (Weaver, 2013).

Thinking about the beyond is a future orientation, where the beyond is a speculative space that messes with linearity. To be against something is to rub against it, to forcefully reject its promise, and perhaps to be willfully avoidant of it. To be beyond something is to return to it from a future and transform the present to ensure the beyond sustains. Being against something always means being for something else; to be beyond something means imagining and enacting alternative visions and spaces of and for the future in the present, not outside of politics but within them. It is the work of this beyond to be against and for something by taking life and death seriously. To be against something is to reach beyond the same something; to be against everything is a radical departure from how things are, towards how things should be (following the words of The Little Brown Dog memorial). Enacting a different future beyond the present is imbued with obligations and duties, but also with endless creative and destructive potential of alternative world building. The work in Part Three of this book is an attempt to take one such building of a world and mess with its very existence within and out-with my attention. What happens to this space when I leave? What happens when I return? Friendship and truth sustain beyond life and beyond death where commonality and specificity must find a way to co-exist.

Uncertainty is a vital part of this orientation towards a future beyond this field/work; beyond the horizon of what and how we know, and who we may become; and beyond embodied finitude (Stanescu, 2012). Rather than presenting solid truths of the way things were and are, instead seeking 'specific material engagements that participate in (re)configuring the world ... making knowledge about specific worldly configurations' (Barad, 2007, 91). Ends can be found located within the world. Their meaningfulness (Kohn, 2013) is thinkable, knowable, speakable, and imaginable in multispecies relational flourishing together. For Wadiwel, following Scarry (1985), uncertainty must be dealt with by bringing others' unknowable experiences into communication with our own: 'the essentially solitary experience of pain makes the ethical question of how we respond to it politically fraught, since we must deal not only with the practical question of what resources should be mobilised to respond to known suffering, but also questions about whether suffering is occurring, how it can be understood and whether it is important enough to respond to' (Wadiwel, 2016, 2).

Living with chickens entails questioning personal, collective, and worldly knowings and doings (Figure 7.3). When truth becomes untruth, uncertainty transforms the world into a strange, bifurcated one, disturbed, and disturbing. The field disperses but remains in my imagining as it was, not as it is. Working with and in the beyond allows me to occupy and keep open this field between the past and the future. Moving further away always returns me closer to the ethical, political, and bodily realities of this work. Something between us will always remain strange, there will always be an unknowability inherent to any relationship, human, animal, other, or self. This unknowability does not absolve us of action but demands protection of the boundaries of difference within the spectral inhabitancies of these differences, to ensure that the diversities we need to sustain the world are inherited, not dispersed to the absolute elsewhere from which the return cannot return and where the beyond is unreachable.

Figure 7.3 Four chickens. Copyright Catherine Oliver 2018.

Summarising Part 3

Part 3 of this book, across Chapters 6 and 7, is located 'after' veganism, and undertakes speculative geographical work on how beyond-human worlds might be lived. Animals have a long history in geographical thought, where animal geography situates animals within assemblages of things and places constituting the multinatural (Lorimer, 2012). Despite its concerns with places, processes and ordering society and environment, animal geography traditionally has not engaged with critiquing the status or human/non-human relations (Castree, 2000). Critical animal geography is determinedly concerned with the problems human exceptionalism poses, where animals are understood as 'subjects of and in spatially uneven practices' (Hobson, 2007, 253). The chicken has a particular place in spatial – even global – configurations.

Changing patterns of human consumption are reconfiguring the Earth's biosphere and the chicken (specifically the broiler) has been proposed as a distinct new morphospecies signal of this (Bennett et al. 2018). With a population of over 22 billion, 'the potential rate of carcass accumulation of chickens is unprecedented in the natural world' (ibid, 7). Even at a slower growth rate, global egg production has increased from 20 million tonnes to over 70 million tonnes per year since 1961. These six chickens that have featured across Chapters 6 and 7 are not *just* chickens, but part of a shifting global ecology. Living with backyard chickens is not only differently navigating the world, disturbing dominant truths, and navigating interspecies relations through re-navigations of somatic and proximal distance and closeness, but also has wider relevance to the ways we might rethink beyond-human violence. Living with chickens entails questioning personal, collective, and worldly knowings and doings. When truth becomes untruth, uncertainty transforms the world into a strange, bifurcated one, disturbed, and disturbing.

Part 3, and living with chickens, should not be dismissed as a hyperlocal concern but as connected with and in relation to changing global circumstances. In 2020, through the Covid-19 pandemic and lockdown, humans were exposed to the threat and consequences of eating animals poses to humanity and public health. Where conversations are taking place on the human abandonments and violences of the pandemic and how society needs to adapt, animals whose abuse is so closely entangled with both Covid-19 and possible future pandemics have also been brought into new spaces. For those of us fortunate enough to have such an experience of the pandemic, the lockdown's slowing of time has opened the possibility to cultivate our domestic space as habitable for other species. During the lockdown in Britain, interest and participation in domestic hen-keeping and rehoming surged for two reasons: people having the time to keep chickens, and the presumption of access to a supply of fresh eggs in the face of supermarket shortages (Mellen, BBC, 2020).

This is not an entirely new phenomenon; there had already been a rise in backyard hen-keeping in Britain, of both ex-laying hens and specialised breeds. Whilst there is clearly a classed element to this – not everyone having a garden in which to keep hens – 5% of the 200,000 hens adopted in the United Kingdom between 2005 and 2012 (Blecha and Leitner, 2014) were in urban London. Given that within urban spaces, chickens and other farmed animals have traditionally been kept distanced from humans, and physically removed from human dwellings (Morin, 2018), this rehoming constitutes a significant shift in the organisation of urban space. Indeed, the rise of animal keeping before and during lockdown perhaps reveals a desire for human reconnection with animals and the natural world. It might be surprising, that a zoonotic disease pandemic has led to an increased desire for increased proximity to other species, but arguably this continues a longstanding and romanticise return to nature (Searle and Turnbull, 2020).

In the face of ethical, environmental, and health challenges, there has been a drastic increase in veganism in the last five years and one in five people have cut down animal meat consumption during the Covid-19 pandemic. The lockdown has to some respects built on and perhaps expedited pre-existing trends in human-animal relationships. Whilst these spatial and behavioural changes are no doubt a good thing in terms of building more equitable and sustainable worlds, Covid-19 opens a series of broader ethical and practical questions around our relationships with other species: How do and should we relate to animals as food providers? What kind of governance is being exerted over non-human animals and their welfare in urban spaces? And, how can we extend these multispecies urban communities beyond just 'beneficial' animals like chickens (Lorimer, 2020) who can comfortably belong in domestic spaces, to other species such as rats, bees, foxes, or slugs all of whom also play vital roles in urban ecosystems? Addressing these questions will be vital in facing up to the key challenges posed both by Covid-19, but also by the key political, ethical, and environmental challenges that both precede and lie dormant in its wake.

Conclusion

One of the difficulties of writing a book that directly and indirectly is about other animals is the careful balancing act between moving beyond the human perspective and considering how animals have mattered to humans and the spatial transformations resulting from this mattering. Animals are already part of our social, political, and cultural worlds. It is no longer the time to bring animals into the geographic discipline, but to consider how rewriting and rethinking geographical theory and practice – and, of course, space itself – is always already multispecies. From this, a different spatial politics emerges. This book traces how this spatiality has been, is, and might be being practised through the practice of veganism as refusal, but also how it allows for building alternative geographical imaginations of multispecies worlds.

What, then, does veganism have to offer to geographical thinking? And how does paying attention to somehow different multispecies worlds allow us to advance our understanding of the spaces within beyond and between us? Across three parts, concerned broadly with historical, contemporary, and emerging spaces of animal activism respectively, the beyond-human geographies and multispecies worlds in this book are not separate entities, but intimately related to one another and to the wider world around them.

Veganism has received little critical attention in geography. Despite its long history, its socio-political force, and its recent surge leading to new global and local geographies, veganism remains marginal to serious geographical consideration. Where veganism has and can be critiqued as upholding whiteness, this book has also upheld parochial Anglo-American dominant perspectives on animal thinking and practice at the expense of global perspectives, largely due to the location of the archive, the research, and myself. By understanding how the narratives of veganism became ones of whiteness and power through the archives of a small group of privileged and powerful dominating the animal movement, we can better resist and challenge them. As the discipline of vegan geographies grows, challenging dominant histories is not enough; there must also be space to find and document diverse global histories.

Contemporary veganism incorporates health, the environment, and compassion for animals. It is a growing spatio-temporal force that seeks to engage with and offer part of a solution to global crises such as climate change, pain, and modern nutrition. Veganism as an organised political movement has done so since its inception, responding to a changing world. In this book, I have attended to some of the ways that caring about animals has been realised in British animal activism and veganism. In Part 1, this is through understanding the histories of animal activism as rooted in friendship as a powerful force. Part 2 contends that contemporary veganism is a refusal and action that is transforming social space, beginning at the scale of the body. Part 3 offers an alternative geographical vision of what multispecies living can look like with a vegan ethic, and how closeness with actual animals changes vegan sentiments.

In Part 1, I consider the dominant narratives of vegan history, which have been forcefully mainstreamed at the expense of work far longer and more radical. The genealogy of Singer, Ryder et al has become so ubiquitous in animal studies that it is difficult to construct ideas outside of these texts. Critical animal studies in particular is being built primarily on the work of these thinkers, to the detriment of earlier and more radical scholars (Adams and Gruen, 2014) and despite their own acknowledgement of the histories they are erasing.

History as community is a spacing and timing of 'us' in, and of which, we must ask politicised questions. Who has defined these histories? Who is invited to be in common? And who is excluded from being part of this community? 'A history bends under its own advance and under its own absence of future ("no future"), unless that history is able to deploy itself towards an unknown opening to the "future to come"'(Nancy, 2014, 1). The centring of the voices of a small group of privileged men has implications for understanding veganism, and its connection to animals and animal activism. Through the friendship of a small group of men, other histories are wiped, ignored, refused, fragmented, and erased.

In Part 2, the re-emergence of animal rights discourses within contemporary veganism reveals how the dominance of Singer et al. persists, especially in relation to the strength of 'truth' narratives. However, this re-emergence of animal rights discourse sits uncomfortably with simultaneous turns to embodiment in contemporary vegan's narratives. The rhetoric of 'rights' is itself dubious, potentially continuing to enforce a hierarchy of beings that asks who deserves which rights and how (Cochrane, 2018). The animal rights movement was never a vegan or anti-speciesist collective, but rather fought predominantly single issues only for some animals, not all (Wrenn and Johnson, 2013).

Part 2 also grapples with navigations of distance (felt, somatic, proximal, imagined – Chapter 4) as vital to contemporary vegan narratives. Veganism is both a socio-spatial refusal (of what vegans don't do/eat/wear) and a renegotiation and transformation of space, filled with potential for building alternative worlds and spaces (Donaldson and Kymlicka, 2015). Narratives of 'truth' in veganism are individual, describing how people decided to become vegan, but also construct shared social bonds within the vegan community through binding shared knowledges and experiences. A shared narrative, or 'truth,' is vital to the collective identity of vegans. Developing and becoming a vegan insider requires not only exposure to new forms of knowledge, but to perform and embody these in visible ways. Paying attention to the embodied elements of vegan knowledge and practice reframes veganism as an ongoing process and renegotiation of body and/in space, both of which 'truth' travels through and transforms.

Part 3 is centred around a multispecies ethnography with six chickens. It attends to the intricacies and complexities of geographical scholarship and analysis might well be overlooking a vital force in thinking about the changing worlds we are in. (Un)thinking human supremacy requires both theoretical and practical interventions and transformations of space and experiments in different kinds of living. Sometimes animals win out over humans; sometimes our worlds are not shareable; sometimes our intentions are not prioritising animals even if we thought they were. Our work and lives with animals must be under constant scrutiny and reflection, in case the violences we seek to end simply change form and reassert themselves.

As animal studies grow, we are developing shared practices and methodologies, informed by and informing our ongoing spatial practices, as we work out the ethical and political implications of different kinds of living together. These new ways of living with the animals already in our lives, such as more equitable practices of walking with dogs or horses (Spannring, 2019), call attention to care and control within changing multispecies worlds (Gillespie, 2019).

Where geographies of the somehow-other-than-human are rooted in and building out of political theories of rights, sociologies of difference, and cultural geographies, these geographies cannot be theorised without attending also to the growth of veganism. The political and ethical implications of this work are relevant across geographical studies concerned with multispecies worlds and futures. Disrupting linearities of time and community requires beyond-human geographies; these departures have long been present in indigenous and majority world perspectives (Todd, 2015). While entanglement scholarship proliferates, it will amount to little, politically and practically, if it continues to rest upon the assumption that the future will look much the same as the past. Veganism is one way into a somehow different future, and demands being taken seriously.

An ending

During the writing of this book, Lacey and Bluebell died. Winnie, Cleo, Primrose, and Olive went through a shift in their social relations and individual temperaments. Maybe they were quieter. They didn't dig under the fence to the freedom of the fields and the ditch anymore. They missed Lacey and Bluebell. Two deaths in quick succession affected their group, their routine, their habits. Over the next year, Cleo, Winnie, Olive, and finally Primrose also died. Each quietly, in their sleep, leaving behind them ever rearticulated and renegotiated relations.

These strange, winged creatures disrupted and disturbed my navigation of the world. Our friendship persists beyond their death, and beyond the present (Derrida, 2005). They are held with me even as the distance between us expands. 'Sometimes we have to do the work even though we don't yet see a glimmer on the horizon that it's actually going to be possible' (Davis, 2016, 29–30). Living with chickens is an intentional choice that articulates a different kind of future: one that brings nonhuman, exploited life into human space differently. Read alongside practices of vegan refusal, these futures take on a deeply political tone that allows us to reconfigure the potential future of multispecies spaces. In becoming close to chickens, we can realise our shared ways of being in world, our commonness, and recognise that we might be able to act in friendship around new truths. Following Nelson and Braun (2017, 233) 'it is not enough to extend the "common" to include non-human capacities; it is equally as important that the common be understood as immanent to concrete arrangements of existence/existents.'

Living with chickens, even in their flightless state, has enabled a different vision of a multispecies world akin to what Hinchcliffe and Whatmore (2006) might have imagined within an urban 'politics of conviviality.' While I researched and wrote this book, I lived in Birmingham. Not far from where the Peregrine falcons that Hinchliffe and Whatmore wrote about have settled in the city centre is a path down onto Birmingham's sprawling canal network that leads in one direction to the Bourn brook, where water voles might dwell. This canal path runs alongside the city's cross city train line, where 'complex assemblages, mutually affecting and affected by their fields of becoming,' of other-than-human species and individuals dwell and 'enfold human and nonhuman mappings' (ibid. 128–9). Where the semi-rurality of Lancashire allowed a sanctuary from the usual human-chicken relations, the urban space of Birmingham offers different forms of, temporally bounded, intimacy with the multispecies urban community.

Between the city centre and the university, there is a stretch of canal which has a path on one side and on the other are trees, shrubs, grasses, flowers – cordoned off from gardens and roads by high metal fences. Walking home along the canal in late Autumn 2019, I saw a heron for the first time in Birmingham (Figure C.1). They were so close I could see the water slide off their waxy feathers. Crouched down, their neck low to the ground. What were they looking at? Surely there can't be fish in that canal? I wonder why this is the first time I have encountered them, when I have been walking these canals for years. Maybe they didn't want to be seen, or maybe something has changed. I don't know, I can't know. I see them often now.

Figure C.1 Heron. Copyright Catherine Oliver 2019.

Weeks later, I saw a different heron, almost blue-grey rather than the other's white-grey, and this one is much smaller (Figure C.2). I stop and crouch on the opposite bank of the canal. The heron is partially hidden by the reeds, standing just in front of what I know is a fox den, where I saw two foxes playing a few weeks ago. The canal cuts between us, separating me from them, and them from me. A cyclist whooshes past and disrupts this world beyond me, beyond them, beyond here.

There's something clumsy about these herons, even though everything about them should be elegant: their stretched legs, long necks, and beautiful shades of white, blue, and grey offer camouflage and splendour. The way they rustle in the bushes on the opposite bank of the canal is as if their

Figure C.2 Heron takes flight. Copyright Catherine Oliver 2019.

attempt to hide has been betrayed by their own stature. Many of the animals I speak to, encounter, and care about in Birmingham are ones I have met on the canals that I walk each day. The canals are a place for city dwelling animals to hunt, play, and sleep.

Where cities are dangerous yet fertile playgrounds for some species, they also offer new forms of living at the same time as obliterating the old ones. This heron, or these herons, stop me each time I see them. I pause, crouch, and fall quiet, hoping for a momentary glimpse into their world, trying not to disturb them even after they have seen me. In the second photograph, the heron had been on the path I was walking, neck bent scanning the canal for food. I stopped about ten metres away and squatted down. All quiet in the middle of the day aside from a train in the distance and children playing at the school half a mile or so away. They saw me, eyeballed me, and fell still too. Both captured in a moment of surprise, the distance held to keep us as strangers.

While the canals are beautiful, bursting with life and offering a brief sanctuary, they are also eerie, dirty, and full of potential dangers. We become prey so easily on these canals. Another world was possible in that elongated moment, until a cyclist dashed through, disturbing our connection. The heron and I become strangers once again. The stranger is a unique sociological category breaching space, distance, and knowledge in their marginality in the ultimate paradox of 'complete liberation and absolute fixation' (McLemore, 1970, 86). The stranger's movement is contingent on my stillness, remaining on the outside, in their beyond. I come close to the heron, we become less strange. The possibility of friendship – of a different world, new 'truths,' shifting temporality – is intimately connected to the stranger. Through the histories, presents, and futures of my work, I was able to approach this heron differently.

'History does not belong primarily to time, nor to succession, nor to causality, but to community, or to being-in-common' (Nancy, 1990, 149). The worlds I have written about in this book – histories of friendship and activism; the contemporary contours of veganism; and a practice of living differently with chickens – are all spaces that strive to understand and realise this community, this 'being-in-common.' The present, then, is 'a kind of project that involves inhabiting a community without guarantees' (Scott, 1999, 95), a sentiment common to the histories, presents, and futures in this book. Without guarantee, it bears repeating that 'we have to do the work even though we don't yet see a glimmer on the horizon that it's actually going to be possible' (Davis, 2016, 29–30). Animals have always been a part of a beyond-human geography. The work of veganism is a fundamental part of continuing to develop a geography that matters, and one that makes a difference beyond the human. Even when it has seemed impossible, humans and animals have – and continue to – build community, negotiate space, and live together in new ways in the face of uncertain futures.

Bibliography

Front matter bibliography

Davis, K. (2009) *Prisoned Chickens, Poisoned Eggs*, Tennessee: Book Publishing Company.

Gregory, J. (2007) *Of Victorians and Vegetarians*, London: IB Tauris.

Hamlett, J. (2019) Britain is a nation of pet lovers and it has the Victorians to thank, *The Conversation*. Available at: https://theconversation.com/britain-is-a-nation-of-pet-lovers-and-it-has-the-victorians-to-thank-121888

Henderson, R.K. (1948) We have been warned!, *The Vegan*. Available at: https://issuu.com/vegan_society/docs/the-vegan-autumn-1949

Kean, H. (1998) Animal rights: Political and social change in Britain since 1800. Reaktion Books: London.

Monson, S. (2005) Earthlings [film]. Burbank, CA. Earthlings.com

Oliver, C. (2018) The women of animal rights, *Archiving Activism*. Available at: https://archivingactivism.wordpress.com/2018/09/03/the-women-of-animal-rights/

Oliver, C. (2020) Beyond-human worlds: Negotiating silence, anger and failure in multispecies research, *Emotion, Space and Society*. https://doi.org/10.1016/j.emospa.2020.100686

Otter, C. (2020) *Diet for a Large Planet: Industrial Britain, Food Systems and World Ecology*, Chicago: University of Chicago Press.

Semple, D. (1945) Health and the soil, *The Vegan*, Spring 1945. Available at: https://issuu.com/vegan_society/docs/the-vegan-news-no.-3-may-1945 8 accessed 15/3/2019

Semple, D. (1947) Live closer to nature, *The Vegan*. Available at: https://issuu.com/vegan_society/docs/the-vegan-autumn-1947 accessed 15/3/2019

Smith, D. (1949) Food for vegan thought, *The Vegan*, Summer Issue. Available at: https://issuu.com/vegan_society/docs/the-vegan-summer-1949

Watson, D. (1944) *The Vegan News*. Available at: https://issuu.com/vegan_society/docs/the_vegan_news_1944 accessed 10/10/2019

Wills, J. (2018) A nation of animal lovers? *The Case for a General Animal Killing Offence in UK Law, King's Law Journal*, 29(3), 407–436.

Introduction bibliography

Adams, C.J. (1994) *Neither Man Nor Beast: Feminism and the Defense of Animals*, London: Bloomsbury.

Adams, C.J. and Gruen, L. (2014) *Ecofeminism*, London: Bloomsbury.

Adams, C.J. (2010) *The Sexual Politics of Meat* (20th Anniversary Edition), New York: Continuum.

Agamben, G. (2009) *What Is an Apparatus?* California: Stanford University Press.

Bekoff, M. (2002) *Minding Animals: Awareness, Emotions, and Heart*, Oxford: Oxford University Press.

Besio, K. (2005) Telling stories to hear autoethnography: Researching women's lives in northern Pakistan, *Gender, Place & Culture*, 12(3), 317–331.

Braidotti, R. (1994) *Nomadic Subjects: Embodiment and Difference in Contemporary Feminist Theory*, New York: Columbia University Press.

Brophy, B. (1971) The rights of animals, in eds. R. Godlovitch, S. Godlovitch and J. Harris, *Animals, Men and Morals: An Inquiry into the Maltreatment of non-Humans*, London: Victor Gollancz.

Castree, N. (2000 [2000]) The production of nature, in eds. E. Sheppard and T.J. Barnes, *A Companion to Economic Geography*, Oxford: Blackwell Publishing, 275–289.

Cherry, E. (2006) Veganism as a cultural movement: A relational approach, *Social Movement Studies*, 5(2), 155–170.

Cochrane, A. (2018) *Sentientist Politics: A Theory of Global Inter-Species Justice*, Oxford: Oxford University Press.

Collard, R.C. and Gillespie, K. (2015) Introduction, in eds. K. Gillespie and R.C. Collard, *Critical Animal Geographies: Politics, Intersections and Hierarchies in a Multispecies World*, London: Routledge.

Daigle, M. (2016) Writing the lives of others: Storytelling and international politics, *Millennium*, 45(1), 25–42.

Deleuze, G. and Guattari, F. (1994) *What Is Philosophy?* New York: Columbia University Press.

Ellis, C., Adams, T.E. and Bochner, A.P. (2011) Autoethnography: An overview, *Historical Social Research/Historische sozialforschung*, 36(4), 273–290.

Figley, C. (2013) *Compassion Fatigue*, New York: Routledge.

Foucault, M. (1997) Friendship as a way of life, *Ethics: Subjectivity and Truth* (1st Edition), Paris: Editions Gallimard, 135–140.

Fraiman, S. (2012) Pussy panic versus liking animals: Tracking gender in animal studies, *Critical Inquiry*, 39(1), 89–115.

Francione, G.L. (2010) Is every campaign a single-issue campaign? Available at: http://www.abolitionistapproach.com/is-every-campaign-a-single-issue-campaign/ accessed 13/03/2019

Giraud, E. (2013) 'Beasts of Burden': Productive tensions between Haraway and radical animal rights activism, *Culture, Theory and Critique*, 54(1), 102–120.

Godlovitch, R. (1971) Animals and morals, *Philosophy*, 46(175), 23–33.

Goodman, M. (2018) The politics of eating bits and bytes, in eds. T. Schneider, K. Eli, C. Dolan and S. Ulijaszek, *Digital Food Activism. Critical Food Politics*, London: Routledge.

Greenebaum, J. and Dexter, B. (2018) Vegan men and hybrid masculinity, *Journal of Gender Studies*, 27(6), 637–648.

Gruen, L. (2015) *Entangled Empathy: An Alternative Ethics for Our Relationships with Animals*, New York: Lantern Books.

Hadley, J. (2015) *Animal Property Rights: A Theory of Habitat Rights for Wild Animals*. Washington DC: Lexington Books.

Haraway, D. (1988) Situated knowledges: The science question in feminism and the privilege of partial perspective, *Feminist Studies*, 14(3), 575–599.

Hardt, M. and Negri, A. (2004) *Multitude*, New York: Penguin Press.

Harrison, R. (1964) *Animal Machines*, London: Vincent Stuart Publishers Ltd.

Hobson, K. (2007) Political animals? On animals as subjects in an enlarged political geography, *Political Geography*, 26(3), 250–267.

Hovorka, A.J. (2012) Women/chickens vs. men/cattle: Insights on gender-species intersectionalty, *Geoforum*, 43(4), 875–884.

Humane Research Council (2014) Study of current and former vegetarians. Available at: https://faunalytics.org/wp-content/uploads/2015/06/Faunalytics_Current-Former-Vegetarians_Full-Report.pdf

Hyndman, J. (2001) The field as here and now, not there and then, *Geographical Review*, 91(1-2), 262–272.

Joy, M. (2009) *Why We Love Dogs, Eat Pigs, and Wear Cows: An Introduction to Carnism*, San Francisco: Conari Press.

Kean, H. (1998) *Animal Rights: Political and Social Change in Britain Since 1800*, London: Reaktion Books.

Kirksey, S.E. and Helmreich, S. (2010) The emergence of multispecies ethnography, *Cultural Anthropology*, 25(4), 545–576.

LiveKindly (2020) Global vegan food market to surpass $31 billion. Available at: https://www.livekindly.co/global-vegan-food-market-surpass-31-billion/

Lorimer, J. (2012) Multinatural geographies for the anthropocene, *Progress in Human Geography*, 36(5), 593–612.

Massey, D. (2005) *For Space*, London: Sage.

Massey, D., Allen, J. and Sarre, P. (1999) *Human Geography Today*, Cambridge: Polity Press.

McDonald, B. (2000) "Once you know something, you can't not know it" an empirical look at becoming vegan, *Society and Animals*, 8(1), 1–23.

McDowell, L. (1997) Women/gender/feminisms: Doing feminist geography, *Journal of Geography in Higher Education*, 21(3), 381–400.

McHugh, S. (2011) *Animal Stories: Narrating Across Species Lines*. Minneapolis, MN: University of Minnesota Press.

Midgley, M. (1983) *Animals and Why They Matter*, Georgia: University of Georgia Press.

Monson, S. (2005) Earthlings [film]. Burbank, CA. Earthlings.com

Morin, K. (2018) *Carceral Space, Prisoners and Animals*, Oxon and New York: Routledge.

Nancy, J.L. (1990 [1990]) Finite history, in ed. D. Carroll, *The States of "Theory": History, Art, and Critical Discourse*, Stanford, California: Stanford University Press, 149–172.

Philo, C. and Wilbert, C. (2000) Animal spaces, beastly places. *Animal Spaces, Beastly Places*, London: Routledge, 15–50.

Povinelli, E. (2018) Horizons and frontiers, late liberal territoriality, and toxic habitats, *E-Flux*, 90. Available at: https://www.e-flux.com/journal/90/191186/horizons-and-frontiers-late-liberal-territoriality-and-toxic-habitats/ accessed 26/11/2019

Probyn, E. (2016) *Eating the Ocean*, Durham: Duke University Press.

Pyke, S.M. (2019) *Animal Visions: Posthumanist Dream Writing*. Springer.

Regan, T. (1983) *The Case for Animal Rights*, California: University of California Press.

ReportBuyer/Statista (2020) Vegan market: Statistics and facts. Available at: https://www.statista.com/topics/3377/vegan-market/

Shaw, I.G.R. (2010) Sites, truths and the logics of worlds: Alain Badiou and human geography, *Transactions of the Institute of British Geographers*, 35(3), 431–442.

Singer, P. (1975) *Animal Liberation: A New Ethics for Our Treatment of Animals*, New York: HarperCollins Publishers.

Srinivasan, K. (2015) Towards a political animal geography? *Political Geography*, 50, 76–78.

Stallwood, K. (2013) *GROWL: Life Lessons, Hard Truths and Bold Strategies from an Animal Advocate*, New York: Lantern Books.

Stallwood, K. (2015) *Brigid Brophy*. Available at: https://kimstallwood.com/2015/10/14/brigid-brophy/ accessed 10/06/2018

Stanescu, J. (2012) Species trouble: Judith Butler, mourning, and the precarious lives of animals, *Hypatia*, 27(3), 567–582.

Stauffer, J. (2015) *Ethical Loneliness: The Injustice of Not Being Heard*, New York: Columbia University Press.

Summers, A. (Curator in Dept of Manuscripts) letter to Ann Payne (Head of Dept Manuscripts); 5/2/1999, The Ryder Papers at the British Library.

Young, S. (2020) Nearly one in four UK food product releases in 2020 were vegan, *The Independent*. Available at: https://www.independent.co.uk/life-style/vegan-food-launch-greggs-sausage-roll-plant-based-veganuary-health-a9287826.html

The Vegan Geographies Collective: Hodge, P., McGregor, A., Narayan, Y., Springer, S. Véron, O. and White, R.J. (2017) Vegan geographies: Ethics beyond violence: Call for papers. Available at: http://www.academia.edu/28821836/Vegan_Geographies_Ethics_Beyond_Violence accessed 12/06/2017

The Vegan Society (2019) *Statistics*. Available at: https://www.vegansociety.com/news/media/statistics

Vaneigem, R. (1983) *The Revolution of Everyday Life* (2nd Edition). PM Press. San Francisco, Los Angeles.

Wadiwel, D. (2016) Fish and pain: The politics of doubt, *Animal Sentience*, 3(31), 1–8.

White, R. (2015 [2015]) Following in the footsteps of Élisée Reclus, 212–230, in eds. R.J. White II and C. Erika, *Anarchism and Animal Liberation: Essays on Complementary Elements of Total Liberation*, North Carolina: McFarland & Company, 212–230.

Wolch, J. and Emel, J. (1995) *Animal Geographies: Place, Politics, and Identity in the Nature-Culture Borderlands*, London: Verso Books.

Wolch, J., Emel, J. and Wilbert, C. (2003) Theme issue on "Bringing the animals Back", *Environment and Planning D: Society and Space*, 13, 631–670.

Wrenn, C.L. (2012) The role of professionalization regarding female exploitation in the nonhuman animal rights movement, *Journal of Gender Studies*, 24(2), 131–146.

Chapter One Bibliography

Adams, C.J. (2010) *The Sexual Politics of Meat* (20th Anniversary Edition), New York: Continuum.

Arcari, P., Probyn-Rapsey, F. and Singer, H. (2020) Where species don't meet: Invisibilized animals, urban nature and city limits. *Environment and Planning E: Nature and Space*, 2514848620939870.

Bauman, Z. (1990) Modernity and ambivalence, *Theory, Culture and Society*, 7(2), 143–169.

Berlant, L. (2011) *Cruel Optimism*, Durham and London: Duke University Press.

Bordo, S. (2004) *Unbearable Weight: Feminism, Western Culture, and the Body*, California: University of California Press.

Braidotti, R. (1994) *Nomadic Subjects: Embodiment and Difference in Contemporary Feminist Theory*, New York: Columbia University Press.

Butler, J. (1990) *Gender Trouble*, New York: Routledge.

Collard, R.C. and Dempsey, J. (2013) Life for Sale? The politics of lively commodities, *Environment and Planning A*, 45(11), 2682–2699.

Cramer, J., Green, C.P. and Walters, L.M. (2011) *Food as Communication: Communication as Food*, New York: Peter Lang.

Cuomo, C. and Gruen, L. (1998) On puppies and pussies: Animals, intimacy and moral distance, in eds. Bat-Ami Bar On & A. Ferguson: *Daring to Be Good: Essays in Feminist Ethico-Politics* (1st Edition), New York: Routledge, 129–142.

Derrida, J. (1996) *Archive Fever: A Freudian Impression*, Chicago: University of Chicago Press.

Derrida, J. (2005) *The Politics of Friendship*, New York: Verso Books.

Farge, A. (2013) *The Allure of the Archives*, London: Yale University Press.

Fitzgerald, A.J. (2010) A social history of the slaughterhouse: From inception to contemporary implications, *Human Ecology Review*, 17(1), 58–69.

Garlick, B. (2015) Not all dogs go to heaven, some go to battersea: Sharing suffering and the 'Brown dog affair', *Social and Cultural Geography*, 16(7), 798–820.

Gillespie, K.A. (2019) For a politicized multispecies ethnography, *Politics and Animals*, 5, 17–32.

Giraud, E. (2019) *What Comes after Entanglement? Activism, Anthropocentrism and an Ethics of Exclusion*, Durham and London: Duke University Press.

Hall, S. (2001) Constituting an archive, *Third Text*, 15(54), 89–92.

Haraway, D. (1984) *A Cyborg Manifesto, Simians, Cyborgs and Women: The Reinvention of Nature*, New York: Routledge.

Ingold, T. (2013) Anthropology beyond humanity, *Suomen Antropologi: Journal of the Finnish Anthropological Society*, 38(3), 5–23.

Irigaray, L. (1996) *I Love to You: Sketch of a Possible Felicity in History*, New York: Routledge.

Joy, M. (2009) *Why We Love Dogs, Eat Pigs, and Wear Cows: An Introduction to Carnism*, San Francisco: Conari Press.

Kirksey, S.E. and Helmreich, S. (2010) The emergence of multispecies ethnography, *Cultural Anthropology*, 25(4), 545–576.

Krebber, A. and Roscher, M. (2019) *Animal Biography: Reframing Animal Lives*, Cham, Switzerland: Palgrave.

Lee, J.A. (2016) Be/longing in the archival body: Eros and the "Endearing" value of material lives, *Archival Science*, 16(1), 33–51.

Lind af Hageby, L. and Schartau, L.K. (2012) *The Shambles of Science: Extracts from the Diary of Two Students of Physiology*, Charleston: Nabu Press.

Lorimer, J. (2014) On auks and awkwardness, *Environmental Humanities*, 4, 195–205.

Massey, D. (2005) *For Space*, London: Sage.

Massumi, B. (2014) *What Animals Teach Us about Politics*, Durham and London: Duke University Press.

Morin, K. (2018) *Carceral Space, Prisoners and Animals*, Oxon and New York: Routledge.

Morris, D. (1999) McLibel: Do-it-yourself justice, *Alternative Law Journal*, 24(1), 269–273.

Nancy, J.L. (1990 [1990]) Finite history, in ed. D. Carroll, *The States of "Theory": History, Art, and Criticial Discourse*, Stanford, California: Stanford University Press, 149–172.

Perlman, F. (2018) *The Machine and Its Discontents: A Fredy Perlman Anthology, Theory and Practice*, Active Distribution.

Pui-Lan, K. (2009) Elisabeth schüsslet fiorenza and postcolonial studies, *Journal of Feminist Studies*, 25(1), 191–197.

Puig de la Bellacasa, M. (2017) *Matters of Care: Speculative Ethics in More than Human Worlds*, Minneapolis, MN: University of Minnesota Press.

Puwar, N. (2004) Space invaders: Race. *Gender and Bodies Out of Place*, New York: Berg.

Regan, T. (1983) *The Case for Animal Rights*, California: University of California Press.

Rich, A. (2003 [1984]) 2003]) Notes towards a politics of location, in eds. R. Lewis and S. Mills, *Feminist Postcolonial Theory: A Reader*, New York: Routledge, 29–42.

RSPCA (2017) Stop live exports international awareness day, RSPCA Insights: The official blog of the RSPCA. Available at: https://blogs.rspca.org.uk/insights/2017/09/13/stop_live_exports_day/#.WlTG62i0OU1

Ryder, R.D. (1983) *Victims of Science: The Use of Animals in Research*, London: National Anti-Vivisection Society.

Ryder, R.D. (2010) Speciesism again: The original leaflet, *Critical Society*, (2), 1–2.

Saunders, T. (2002) *Baiting the Trap: One Man's Secret Battle to Save Our Wildlife*, New York: Pocket Books.

Scarry, E. (1985) *The Body in Pain: The Making and Unmaking of the World*, Oxford: Oxford University Press.

Singer, P. (1975) *Animal Liberation: A New Ethics for Our Treatment of Animals*, New York: HarperCollins Publishers.

Spivak, G.C. (1988) Can the subaltern speak? in ed. R.C. Morris, *Can the Subaltern Speak? Reflections on the History of an Idea*, New York: Columbia University Press.

Springer, S. (2011) Violence sits in places? Cultural practice, neoliberal rationalism, and virulent imaginative geographies, *Political Geography*, 30(20), 90–98.

Stoler, A.L. (2009) *Along the Archival Grain: Epistemic Anxieties and Colonial Common Sense*, New Jersey: Princeton University Press.

Sutton, Z. (2020) Researching towards a critically posthumanist future: On the political "doing" of critical research for companion animal liberation, *International Journal of Social Policy*, Vol. ahead-of-print No. ahead-of-print.

Taylor, N. and Signal, T.D. (2009) Pet, pest, profit: Isolating differences in attitudes towards the treatment of animals, *Anthrozoös*, 2, 129–135.

Twine, R. (2014) Intersectional disgust? Animals and (eco)feminism, *Feminism & Psychology*, 20(3), 397–406.

Wadiwel, D. (2018) Chicken harvesting machines: Animal labor, resistance, and the time of production, *The South Atlantic Quarterly*, 117(3), 527–549.

Weaver, H. (2013) Becoming in kind: Race, class, gender, and nation in cultures of dog rescue and dogfighting, *American Quarterly*, 65(3), 689–709.

White, R. (2015 [2015]) Following in the footsteps of Élisée Reclus, 212–230, in eds. A.J. Nocella II, R.J. White and C. Erika, *Anarchism and Animal Liberation: Essays on Complementary Elements of Total Liberation*, North Carolina: McFarland & Company, 212–230.

Wolfson, D.J. (1999) McLibel, *Animal Law*, 5, 21–60.

Wrenn, C.L. (2019) *Animal Rights in the Age of non-Profits*. Ann Arbor, MI: University of Michigan Press.

Yancy, G. (2005) Whiteness and the return of the Black body, *Journal of Speculative Philosophy*, 19(4), 215–241.

Chapter Two Bibliography

Adams, C.J. (1975) The oedible complex: Feminism and vegetarianism, in eds. G. Covina and L. Galana, *The Lesbian Reader*, Oakland, CA: Amazon.

Adams, C.J. (1994) *Neither Man Nor Beast: Feminism and the Defense of Animals*, London: Bloomsbury.

Adams, C.J. and Gruen, L. (2014) *Ecofeminism*, London: Bloomsbury.

Beers, D.L. (2006) *For the Prevention of Cruelty: The History and Legacy of Animal Rights Activism in the United States*, Ohio: Ohio University press.

Bekoff, M. (2002) *Minding Animals: Awareness, Emotions, and Heart*, Oxford: Oxford University Press.

Bentham, J. ([1780] 1982) *An Introduction to the Principles of Morals and Legislation*, edited by J.H. Burns and H.L.A. Hart, London: Methuen.

Best, S. and Nocella, A.J. II (2004) Behind the mask: Uncovering the animal liberation front. *Terrorists or Freedom Fighters?: Reflections on the Liberation of Animals*, New York: Lantern Books, 9–64.

Bourke, D. (2019) The animal rights movement and the law: Engagement, co-option and resistance, [PhD thesis], The University of Auckland, New Zealand. Available at: https://researchspace.auckland.ac.nz/handle/2292/47738?fbclid=IwAR1qHaB-ZQMxnlvlOaJTcmyV8JeZkYVmOAR6eXfZwbeMizCZXlvmV4G_DoU

Boyce Davies, C. (1999) Beyond unicentricity: Transcultural black presences, *Research in African Literatures*, 30(2), 96–109.

Brophy, B. (1965) The rights of animals, *The Sunday Times*, October 10, 1965. Accessed at The British Library, the Ryder Papers, January 2017.

Brophy, B. (1971) The rights of animals, in eds. R. Godlovitch, S. Godlovitch and J. Harris, *Animals, Men and Morals: An Inquiry into the Maltreatment of non-Humans*, London: Victor Gollancz.

Craddock, E. (2019) Doing 'enough' of the 'right' thing: The gendered dimension of the 'ideal activist' identity and its negative emotional consequences, *Social Movement Studies*, 18(2), 137–153.

Cross, L.J. (1949) Search of Veganism-1, *The Vegan Summer 1949*, Vol V. No. 2. Available at: https://issuu.com/vegan_society/docs/the-vegan-summer-1949

Davis, K. (1989) Mixing without pain, *Between the Species*, 5(1), 33–37.

De Leeuw, S. (2012) Alice through the looking glass: Emotion, personal connection, and reading colonial archives along the grain, *Journal of Historical Geography*, 38(3), 273–281.

Derrida, J. (1996) *Archive Fever: A Freudian Impression*, Chicago: University of Chicago Press.

Derrida, J. (2002) *The Animal that Therefore I Am*, New York: Fordham University Press.

Derrida, J. (2005) *The Politics of Friendship*, New York: Verso Books.

Farge, A. (2013) *The Allure of the Archives*, London: Yale University Press.

Fleckner, J. (1991) "Dear Mary Jane": Some reflections on being an archivist, *The American Archivist*, 54(1), 8–13.

Fraiman, S. (2012) Pussy panic versus liking animals: Tracking gender in animal studies, *Critical Inquiry*, 39(1), 89–115.

Garner, R. (2004) *Animals, Politics and Morality*, Manchester: Manchester University Press.

Garner, R. (2019) The Oxford Group Blog I: The origins of a friendship group. Available at: https://www2.le.ac.uk/departments/politics/research/research-projects/the-psychogeography-of-the-oxford-group/the-oxford-group-blog-i-the-origins-of-a-friendship-group accessed 21/01/2020

Garner, R. and Okuleye, Y. (2020) *The Oxford Group and the Emergence of Animal Rights: An Intellectual History*, Oxford: Oxford University Press.

Godlovitch, R. (1971) Animals and morals, *Philosophy*, 46(175), 23–33.

Godlovitch, R., Godlovitch, S. and Harris, J. (1971) *Animals, Men and Morals: An Inquiry into the Maltreatment of non-Humans*, London: Victor Gollancz.

Grever, M. (1997) The pantheon of feminist culture: Women's movements and the organization of memory, *Gender & History*, 9(2), 364–374.

Hall, S. (2001) Constituting an archive, *Third Text*, 15(54), 89–92.

Harper, A.B. (2010) *Sistah Vegan*, New York: Lantern Books.

Harper, A.B. (2013) Vegan consciousness and the commodity chain: On the neoliberal, Afrocentric, and decolonial politics of "cruelty-free" [PhD thesis]. Available at: http://sistahvegan.com/wp-content/uploads/2013/03/harperpdftelfordupdates.pdf, accessed 07/01/2020

Harrison, R. (1964) *Animal Machines*, London: Vincent Stuart Publishers Ltd.

Hayes Edwards, B. (2012) The taste of the archive, *Callaloo*, 35(4), 955–972.

Joon-Ho (2017) [film] *Okja*. Netflix.

Kean, H. (1998) *Animal Rights: Political and Social Change in Britain Since 1800*, London: Reaktion Books.

Latour, B. (1993) *We Have Never Been Modern*, Cambridge, MA: Harvard University Press.

McCarthy, J. and Dekoster, S. (2020) Nearly one in four in U.S. have cut back on eating meat, *Gallup*. Available at: https://news.gallup.com/poll/282779/nearly-one-four-cut-back-eating-meat.aspx accessed 23/02/2020

Morris, A. and Oliver, C. (2016) Kale and chia seed smoothie anyone? Instagram's clean-eating trend, classism and the misrepresentation of veganism, *Feminist academic collective*. Available at: https://feministacademiccollective.com/2016/08/24/kale-and-chia-seed-smoothie-anyone-instagrams-clean-eating-trend-classism-and-the-misrepresentation-of-veganism/

Nancy, J.L. (1990 [1990]) Finite history, in ed. D. Carroll, *The States of "Theory": History, Art, and Critical Discourse*, Stanford, California: Stanford University Press, 149–172.

Parkinson, C. (2018) Animal bodies and embodied visuality, *Antennae: The Journal of Nature in Visual Culture*, 46, 51—64.

Phelps, N. (2007) *The Longest Struggle: Animal Advocacy from Pythagoras to PETA*, Cheltenham: Lantern Books.

Preece, R. (2009). *Sins of the Flesh: A History of Ethical Vegetarian Thought.* Vancouver, Canada: UBC Press

Probyn-Rapsey, F., O'Sullivan, S. and Watt, Y. (2019) Pussy Panic and glass elevators: How gender is shaping the field of animal studies, *Australian Feminist Studies*, 34(100), 198–215.

Regan, T. (1983) *The Case for Animal Rights*, California: University of California Press.

Ritvo, H. (1984) Plus ca change: Anti-vivisection then and now, *Science, Technology, & Human Values*, 9(2), 57–66.

Roscher, M. (2009) *A Kingdom for the Animals: The History of the British Animal Rights Movement*, Marburg: Oberhessische Press.

Ruether, R.R. (1975) *New Woman, New Earth: Sexist Ideologies and Human Liberation. Front Cover.* New York: Seabury Press.

Ryder, R.D. (1989) *Animal Revolution: Changing Attitudes Towards Speciesism*, London and Oxford: Bloomsbury.

Ryder, R.D. (1998) Interview recorded by Melanie Oxley, 16-09-1998, interviewee's home, Tape 1, Shelfmark C894/02, 'Animal Welfare Activists' Collection, The British Library.

Ryder, R.D. (2010) Speciesism again: The original leaflet, *Critical Society*, (2), 1–2

Sexton, A. (2004) *A Self-Portrait in Letters* (Reprint Edition), Massachusetts: Mariner Books/Houghton Mifflin Company.

Singer, P. (1972) Review of 'Animals, Men and Morals', The New York Review of Books.

Singer, P. (1975) *Animal Liberation: A New Ethics for Our Treatment of Animals*, New York: HarperCollins Publishers.

Singer, P. (1982) The Oxford vegetarians: A personal account, *International Journal for the Study of Animal Problems*, 3(1), 6–9.

Singer, P. and Regan, T. (1976) *Animal Rights and Human Obligations*, Hoboken, NJ: Prentice Hall.

Solis, G. (2018) Reflections on archives of violence and transformative justice. Available at: https://medium.com/community-archives/reflections-on-archives-of-violence-and-transformative-justice-87e813f310fe accessed 18/6/2019

Stallwood, K. and McKibbin, P. (2019) We are all animal lovers, *Sentient Media.* Available at: https://sentientmedia.org/we-are-all-animal-lovers/ accessed 01/12/2019

Steiner, G. (2005). *Anthropocentrism and Its Discontents: The Moral Status of Animals in the History of Western Philosophy*, University of Pittsburgh Press.

Stewart, K. and Cole, M. (2014) *Our Children and Other Animals: The Cultural Construction of Human-Animal Relations in Childhood*, London: Routledge.

Stoler, A.L. (2009) *Along the Archival Grain: Epistemic Anxieties and Colonial Common Sense*, New Jersey: Princeton University Press.

Summers, A. (Curator in Dept of Manuscripts) letter to Ann Payne (Head of Dept Manuscripts); 5/2/1999, The Ryder Papers at the British Library.

Villanueva, G. (2016) 'The Bible' of the animal movement: Peter singer and animal liberation, 1970-1976, *History Australia*, 13(3), 399–414.

Wrenn, C.L. (2019) *Animal Rights in the Age of non-Profits*, Ann Arbor, MI: University of Michigan Press.

Wright, L. (2020) Vegan studies. Available at: https://medium.com/@laurawright_76468/vegan-studies-920f25e6f274 accessed 20/01/2020

Wynter, S. and McKittrick, K. (2015) Unparalleled catastrophe for our species? Or, to give humanness a different future: Conversations, in ed. K. McKittrick, *Sylvia Wynter: On Being Human as Praxis*, Durham and London: Duke University Press.

Chapter Three Bibliography

Adams, C.J. (2010) *The Sexual Politics of Meat* (20th Anniversary Edition), New York: Continuum.

Badhwar, N.K. (2003 [2003]) Love, in ed. H. LaFollette, *Practical Ethics*, Oxford: Oxford University Press, 42–69.

Badiou, A. (2009) *In Praise of Love*, London: Serpent's Tail.

Bell, G.G. and Zaheer, A. (2007) Geography, networks, and knowledge flow, *Organization Science*, 18(6), 955–972.

Bissell, D. and Gorman-Murray, A. (2019) Disoriented geographies: Undoing relations, encountering limits, *Transactions of the Institute of British Geographers*, 44, 707–720.

Braidotti, R. (1994) *Nomadic Subjects: Embodiment and Difference in Contemporary Feminist Theory*, New York: Columbia University Press.

Bunnell, T., Yea, S., Peake, L., Skelton, T. and Smith., M. (2012) Geographies of friendships, *Progress in Human Geography*, 36(4): 490–507. doi: 10.1177/0309132511426606

Casciani, D. (2014) The undercover cop, his lover, and their son, *BBC*. Available at: https://www.bbc.co.uk/news/magazine-29743857

Cifor, M. (2016) Affecting relations: Introducing affect theory to archival discourse, *Archival Science*, 16(1), 7–31.

Cohn, J. and Wilbur, S. (2003) What's Wrong with Postanarchism? Available at: https://theanarchistlibrary.org/library/jesse-cohn-and-shawn-wilbur-what-s-wrong-with-postanarchism

Dancey-Downs, K. (2018) Exposing the spy who loved me. Available at: https://uk.lush.com/article/exposing-spy-who-loved-me

Davis, A. (2016) In discussion with Boggs, G.L. *Angela Davis on Veganism as Part of a Revolutionary Perspective*. Available at: https://www.filmsforaction.org/watch/angela-davis-on-veganism-as-part-of-a-revolutionary-perspective/ accessed 30/10/2019

Derrida, J. (2005) *The Politics of Friendship*, New York: Verso Books.

Duranti, L. (1997) The archival bond, *Archives and Museum Informatics*, 11(3-4), 213–218.

Emmerson, P. (2019) From coping to carrying on: A pragmatic laughter between life and death, *Transactions of the Institute of British Geographers*, 44(1), 141–154.

Foucault, M. (1997) Friendship as a way of life, *Ethics: Subjectivity and Truth* (1st Edition), 135–140.

Fraiman, S. (2012) Pussy panic versus liking animals: Tracking gender in animal studies, *Critical Inquiry*, 39(1), 89–115.

Gaarder, E. (2011) *Women and the Animal Rights Movement*, New Jersey: Rutgers University Press.

Garlick, B. (2015) Not all dogs go to heaven, some go to battersea: Sharing suffering and the 'Brown dog affair', *Social and Cultural Geography*, 16(7), 798–820.

Gilligan, C. (1982) *In a Different Voice*, Cambridge: Harvard University Press.

Goldman, J. (2010) Who let the dogs Out? Samuel Johnson, Thomas Carlyle, Virginia Woolf and the Little Brown dog. *Virginia Woolf's Bloomsbury, Volume 2*, London: Palgrave Macmillan, 46–65.

Hall, S.M. and Jayne, M. (2016) Make, mend and befriend: Geographies of austerity, crafting and friendship in contemporary cultures of dressmaking in the UK, *Gender, Place and Culture*, 23(2), 216–234.

Haraway, D. (2003) *The Companion Species Manifesto: Dogs, People and Significant Otherness*, Chicago: University of Chicago Press.

Harper, A.B. (2010) *Sistah Vegan*, New York: Lantern Books.

Helm, B. 2017. Friendship, in ed. N. Zalta, *The Stanford Encyclopaedia of Philosophy* (Fall 2017 ed.), Stanford: Metaphysics Research Lab, Stanford University.

Hodgson, N. (2016) The researcher and the studier: On stress, tiredness and homelessness in the university, *Journal of Philosophy of Education*, 50(1), 37–48.

hooks, bell (2000) *All About Love: New Visions*, New York: HarperCollins Publishers.

Kean, H. (2003) An exploration of the sculptures of greyfriars bobby, Edinburgh, Scotland, and the Brown dog, battersea, South London, England, *Society & Animals*, 11(4), 353–373.

Keller, S. (2000) How do I love thee? Let me count the properties, *American Philosophical Quarterly*, 37(2), 163–173.

Kern, L. (2019) *The Feminist City*, London: Verso Books.

Kirsch, G.E. (2005) Friendship, friendliness, and feminist fieldwork, *Signs: Journal of Women in Culture and Society*, 30(4), 2163–2172.

Ko, A. and Ko., S. (2017) *Aphro-Ism: Essays on Pop Culture, Feminism, and Black Veganism from Two Sisters*, New York: Lantern Books.

Levine, C. (2012) The tyranny of tyranny. *Quiet Rumours: An Anarcha-Feminist Reader*, Edinburgh: AK Press Ltd, 77–83.

Lorimer, H. (2003) Telling small stories: Spaces of knowledge and the practice of geography, *Transactions of the Institute of British Geographers*, 28(2), 197–217.

Massey, D. (2005) *For Space*, London: Sage.

McCormack, D.P. (2014) Atmospheric things and circumstantial excursions, *Cultural Geographies*, 21(4), 605–625.

Mills, S. (2013) Cultural–historical geographies of the archive: Fragments, objects and ghosts, *Geography Compass*, 7(10), 701–713.

Montaigne, M. (2004 [1580]) *On Friendship*, London: Penguin Books.

Morin, K. (2018) *Carceral Space, Prisoners and Animals*, Oxon and New York: Routledge.

Nietzsche, F. (1996) Human, all too human, in ed. R.J. Holingdale, *A Book for Free Spirits*, Cambridge: Cambridge University Press.

OED (2019) Friendship, n. Available at: https://www.oed.com/view/Entry/74661?redirectedFrom=friendship#eid accessed 24/10/2019

Oliver, C. (2018) *Archiving Activism: The Animal Guide*. Available at: https://archivingactivism.com/ accessed 10/10/2018

Ono-George, M. (2019) "Power in the telling": Community-engaged histories of black Britain. Available at: http://www.historyworkshop.org.uk/power-in-the-telling/ accessed 19/11/2019

Perlman, F. (2018) *The Machine and Its Discontents: A Fredy Perlman Anthology, Theory and Practice*, Active Distribution.

Rose, S. and Roades, L. (1987) Feminism and Women's friendships, *Psychology of Women Quarterly* 11, 243–254

Scarry, E. (1985) *The Body in Pain: The Making and Unmaking of the World*, Oxford: Oxford University Press.

Simmel, G. (2008) The stranger. In eds. T. Oakes and P.L. Price, *The Cultural Geography Reader*, Routledge, 323–327.

Sokolowski, R. (2002) Phenomenology of friendship, *The Review of Metaphysics*, 55(3), 451–470.

SSPCA (1970) *Annual Pictorial Review*, 27 [The Ryder Papers].

Stewart, K. (2013) *Ordinary Affects*, Durham and London: Duke University Press.

Stroud, S. (2006) Epistemic partiality in friendship, *Ethics*, 116(3), 498–524.

Taylor, S. (2015) Investigation into links between Special Demonstration Squad and Home Office. Available at: https://www.bl.uk/britishlibrary/~/media/bl/global/social-welfare/pdfs/non-secure/i/n/v/investigation-into-links-between-special-demonstration-squad-and-home-office.pdf accessed 30/09/2019

Wadiwel, D. (2016) Fish and pain: The politics of doubt, *Animal Sentience*, 3(31), 1–8.

Webb, D. (2003) On friendship: Derrida, Foucault, and the practice of becoming, *Research in Phenomenology*, 33(1), 119–140.

Chapter Four Bibliography

Adams, C.J. (1994) *Neither Man Nor Beast: Feminism and the Defense of Animals*, London: Bloomsbury.

Boisseau, W. (2015 [2015]) "A wider vision": Coercion, solidarity and animal liberation, in eds. A.J. Nocella III, R.J. White and E. Cudworth, *Anarchism and Animal Liberation: Essays on Complementary Elements of Total Liberation*, North Carolina: McFarland & Company, 141–160.

Brinkmann, J. (2004) Looking at consumer behavior in a moral perspective, *Journal of Business Ethics*, 51(2), 129–141.

Cochrane, A., Garner, R. and O'Sullivan, S. (2018) Animal ethics and the political, *Critical Review of International Social and Political Philosophy*, 21(2), 261–277.

Cole, M. and Morgan, K. (2011) Vegaphobia: Derogatory discourses of veganism and the reproduction of speciesism in UK national newspapers, *The British Journal of Sociology*, 62(1), 134–153.

Craddock, E. (2019) Doing 'enough' of the 'right' thing: The gendered dimension of the 'ideal activist' identity and its negative emotional consequences, *Social Movement Studies*, 18(2), 137–153.

Donald, D. (2019) *Women Against Cruelty: Protection of Animals in Nineteenth-Century Britain*, Manchester: Manchester University Press.

Donaldson, S. and Kymlicka, W. (2011) *Zoopolis: A Political Theory of Animal Rights*, Oxford: Oxford University Press.

Foucault, M. (1997) Friendship as a way of life, *Ethics: Subjectivity and Truth* (1st Edition), 135–140.

Freeman, J. (2012) The tyranny of structurelessness. *Quiet Rumours: An Anarcha-Feminist Reader*, Edinburgh: AK Press Ltd, 77–83.

Gelderloos, P. (2008) Veganism is a consumer activity, *The Anarchist Library*. Available at: http://theanarchistlibrary.org/library/peter-gelderloos-veganism-is-a-consumer-activity

Kean, H. (1998) *Animal Rights: Political and Social Change in Britain Since 1800*, London: Reaktion Books.

Kern, L. (2019) *The Feminist City*, London: Verso Books.

Legal Information Institute (2019) United States code: Title 18, 43. Force, violence, and threats involving animal enterprises, Cornell University Law School.

Malizia, M. (2014) Autobiografias de gerações feministas: Algumas reflexões sobre narrativas de histórias feministas, *Ex aequo*, (30), 23–37.

Matthee, D.D. (2004) Towards an emotional geography of eating practices: An exploration of the food rituals of women of colour working on farms in the Western Cape, *Gender, Place & Culture*, 11(3), 437–443.

McDonald, B. (2000) "Once you know something, you can't not know it" an empirical look at becoming vegan, *Society and Animals*, 8(1), 1–23.

Morris, A. and Oliver, C. (2016) Kale and chia seed smoothie anyone? Instagram's clean-eating trend, classism and the misrepresentation of veganism, *Feminist academic collective*. Available at: https://feministacademiccollective.com/2016/08/24/kale-and-chia-seed-smoothie-anyone-instagrams-clean-eating-trend-classism-and-the-misrepresentation-of-veganism/

O'Brien, K., Selboe, E. and Hayward, B. (2018) Exploring youth activism on climate change: Dutiful, disruptive, and dangerous dissent, *Ecology and Society*, 23(3), 4–.

Oliver, C. (2018) *Archiving Activism: The Animal Guide*. Available at: https://archivingactivism.com/ accessed 10/10/2018

Oliver, C. (2020) Beyond-human worlds: Negotiating silence, anger & failure in multispecies research, *Emotion, Space and Society*, https://doi.org/10.1016/j.emospa.2020.100686

Pfeffer, M.J. and Parson, S. (2015 [2015]) Industrial society is both the fabrication department and the kill floor: Total liberation, green anarchism and the violence of industrialism, in eds. A.J. Nocella III, R.J. White and E. Cudworth, *Anarchism and Animal Liberation: Essays on Complementary Elements of Total Liberation*, North Carolina: McFarland & Company, 141–160.

Pottinger, L. (2017) Planting the seeds of a quiet activism, *Area*, 49(2), 215–222.

Ramírez, M.M. (2014) The elusive inclusive: Black food geographies and racialized food spaces, *Antipode*, 47(3), 748–769.

Smith, D.M. (1997) Back to the good life: Towards an enlarged conception of social justice, *Environment and Planning D: Society and Space*, 15(1), 19–35.

Socha, K. (2013 [2013]) The "dreaded comparisons" and speciesism: Levelling the hierarchy of suffering, in eds. K. Socha and S. Blum, *Confronting Animal Exploitation: Grassroots Essays on Liberation and Veganism*, North Carolina and London: McFarland & Company, 223–240.

Stallwood, K. (2013) *GROWL: Life Lessons, Hard Truths and Bold Strategies from an Animal Advocate*, New York: Lantern Books.

Stallwood, K. and McKibbin, P. (2019) We are all animal lovers, *Sentient Media*. Available at: https://sentientmedia.org/we-are-all-animal-lovers/ accessed 01/12/2019

Stanescu, J. (2012) Species trouble: Judith Butler, mourning, and the precarious lives of animals, *Hypatia*, 27(3), 567–582.

White, R. (2015 [2015]) Following in the footsteps of Élisée Reclus, 212-230, in eds. A.J. Nocella II, R.J. White and E. Cudworth, *Anarchism and Animal Liberation: Essays on Complementary Elements of Total Liberation*, North Carolina: McFarland & Company, 212–230.

Woodman, C. (2018) *Spycops in Context: Counter-Subversion, Deep Dissent and the Logic of Political Policing*, London: Centre for Crime and Justice Studies.

Yates, R. (2020) The battle for the heart of veganism, *Vegan Sociology Conference 2020*. Available at: https://www.youtube.com/watch?v=RlGc5WvaoiQ&ab_channel=VeganSociology

Chapter Five Bibliography

Adams, C.J. (2010) *The Sexual Politics of Meat* (20th Anniversary Edition), New York: Continuum.

Adams, C.J. (2018) *Burger*, New York: Bloomsbury.

Anzaldúa, G. (1987) *Borderlands/La Frontera: The New Mestiza*, California: Aunt Lute Books.

Butler, J. (1990) *Gender Trouble*, New York: Routledge.

Craddock, E. (2019) Doing 'enough' of the 'right' thing: The gendered dimension of the 'ideal activist' identity and its negative emotional consequences, *Social Movement Studies*, 18(2), 137–153.

Fienup-Riordan, A. (2000 [2000]) An anthropologist reassess her methods, in eds. A. Fienup-Riordan and J. Active, *Hunting Tradition in a Changing World: Yup'ik Lives in Alaska Today*, New Brunswick: Rutgers University Press, 29–57.

Fitzgerald, A.J. (2010) A social history of the slaughterhouse: From inception to contemporary implications, *Human Ecology Review*, 17(1), 58–69.

Goffman, E. (1956) *The Presentation Self in Everyday Life*, New York: Random House.

Haraway, D. (2016) *Staying with the Trouble: Making Kin in the Chthulucene*, Durham and London: Duke University Press.

Harper, A.B. (2010) *Sistah Vegan*, New York: Lantern Books.

Ingold, T. (2011) *Being Alive: Essays on Movement, Knowledge and Description*, New York: Routledge.

Johnson, A. (2017) Getting comfortable to feel at home: Clothing practices of black Muslim women in Britain, *Gender, Place & Culture*, 24(2), 274–287.

Ko, A. and Ko., S. (2017) *Aphro-Ism: Essays on Pop Culture, Feminism, and Black Veganism from Two Sisters*, New York: Lantern Books.

Kundera, M. (1985) *The Unbearable Lightness of Being*, New York: Harper & Row.

Lockwood, A. (2018) Bodily encounter, bearing witness and the engaged activism of the global save movement, *The Animal Studies Journal*, 7(1), 104—126.

Mies, M. and Shiva, V. (2014) *Ecofeminism*, London: Zed Books.

Moraga, C. and Anzaldúa, G. (1981) *This Bridge Called My Back: Writings by Radical Women of Colour* (4th Edition), Albany: State University of New York Press.

Morin, K. (2018) *Carceral Space, Prisoners and Animals*, Oxon and New York: Routledge.

Pfeffer, R. (1965) Eternal recurrence in Nietzsche's philosophy, *The Review of Metaphysics*, 19(2), 276–300.

Povinelli, E. (2011) *Economies of Abandonment*, Durham and London: Duke University Press.

Povinelli, E. (2018) Horizons and frontiers, late liberal territoriality, and toxic habitats, *E-Flux*, 90. Available at: https://www.e-flux.com/journal/90/191186/horizons-and-frontiers-late-liberal-territoriality-and-toxic-habitats/ accessed 26/11/2019

Ryder, R.D. (1998) Interview recorded by Melanie Oxley, 16-09-1998, interviewee's home, Tape 1, Shelfmark C894/02, 'Animal Welfare Activists' Collection, The British Library.

Scarry, E. (1985) *The Body in Pain: The Making and Unmaking of the World*, Oxford: Oxford University Press.

Spiegel, M. (1997) *The Dreaded Comparison: Human and Animal Slavery*, New York: Mirror Books.

Springer, S. and Le Billon, P. (2016) Violence and space: An introduction to the geographies of violence, *Political geography*, 52, 1—3.

Stone, J. (2019) Campaigners rally against EU 'veggie burger' name ban, *The Independent*. Available at: https://www.independent.co.uk/news/world/europe/eu-ban-veggie-burgers-campaign-disks-name-change-european-parliament-a8912116.html

Todd, Z. (2015) Indigenizing the Anthropocene. *Art in the Anthropocene: Encounters among Aesthetics, Politics, Environments and Epistemologies*, 241—254.

Wadiwel, D. (2016) Fish and pain: The politics of doubt, *Animal Sentience*, 3(31), 1–8.

White, R. (2015 [2015]) Following in the footsteps of Élisée Reclus, 212–230, in eds. A.J. Nocella II, R.J. White and E. Cudworth, *Anarchism and Animal Liberation: Essays on Complementary Elements of Total Liberation*, North Carolina: McFarland & Company, 212–230.

Widdows, H. (2018) *Perfect Me: Beauty as an Ethical Ideal*, New Jersey: Princeton University Press.

Yancy, G. (2005) Whiteness and the return of the Black body, *Journal of Speculative Philosophy*, 19(4), 215–241.

Chapter Six Bibliography

Ahmed, S. (2000) *Strange Encounters: Embodied Others in Post-Coloniality*, Oxon and New York: Routledge.

Ahmed, S. (2017) *Living a Feminist Life*, Durham and London: Duke University Press.

Ahmed, S. (2019) Why Complain? Available at: https://feministkilljoys.com/2019/07/22/why-complain/ accessed 22/07/2019

Barua, M., White, T. and Nally, D. (2020) Rescaling the Metabolic, *CRASSH*. Available at: http://www.crassh.cam.ac.uk/blog/post/rescaling-the-metabolic

Beldo, L. (2017) Metabolic labor: Broiler chickens and the exploitation of vitality, *Environmental Humanities*, 9(1), 108–128.

Bissell, D. and Gorman-Murray, A. (2019) Disoriented geographies: Undoing relations, encountering limits, *Transactions of the Institute of British Geographers*, 44, 707—720.

Capps, A. (2014) Eggs: What are you really eating? *Free from harm*. Available at: https://freefromharm.org/eggs-what-are-you-really-eating/

Cavalieri, P. (2009) *The Death of the Animal: A Dialogue*, New York: Columbia University Press.

Cochrane, A. (2018) *Sentientist Politics: A Theory of Global Inter-Species Justice*, Oxford: Oxford University Press.

Davis, K. (2009) *Prisoned Chickens, Poisoned Eggs*, Tennessee: Book Publishing Company.

DEFRA (2019) Code of practice for the welfare of laying hens and pullets. Available at: https://www.gov.uk/government/publications/poultry-on-farm-welfare/ poultry-welfare-recommendations#poultry-housing-environment

DeMello, M. (2012) *Teaching the Animal: Human-Animal Studies Across the Disciplines*, New York: Lantern Books.

Dinker, K.G. and Pedersen, H. (2016 [2016]) Critical animal pedagogies: Re-learning our relations with animal others, in eds. H.E. Lees and N. Noddings, *The Pal-Grave International Handbook of Alternative Education*, London: Palgrave MacMillan, 415–430.

Donaldson, S. and Kymlicka, W. (2015) Farmed animal sanctuaries: The heart of the movement, *Politics and Animals*, 1(1), 50–74.

Gillespie, K.A. (2019) For a politicized multispecies ethnography, *Politics and Animals*, 5, 17–32.

Giraud, E. (2019) *What comes after Entanglement? Activism, Anthropocentrism and an Ethics of Exclusion*, Durham and London: Duke University Press.

Godfrey-Smith, P. (2016) *Other Minds: The Octopus, the Sea and the Evolution of Intelligent Life*, London: William Collins.

Gruen, L. (2015) *Entangled Empathy: An Alternative Ethics for Our Relationships with Animals*, New York: Lantern Books.

Hamilton, L. and Taylor, N. (2017) *Ethnography after Hu-Manism: Power, Politics and Method in Multi-Species Research*, London: Palgrave Macmillan.

Haraway, D. (2008) *When Species Meet*, Minneapolis, MN: University of Minnesota Press.

Hepperman, C. (2012) *City Chickens*, New York: Houghton Mifflin.

Hovorka, A.J. (2012) Women/chickens vs men/cattle: Insights on gender-species intersectionalty, *Geoforum*, 43(4), 875–884.

Irigaray, L. (1996) *I Love to You: Sketch of a Possible Felicity in History*, New York: Routledge.

Jordan, J. (2001) Why friends Shouldn't let friends be eaten: An argument for vegetarianism, *Social Theory and Practice*, 27(2), 309–322.

Joy, M. (2009) *Why We Love Dogs, Eat Pigs, and Wear Cows: An Introduction to Carnism*, San Francisco: Conari Press.

Kohn, E. (2013) The living thought. *How Forests Think: Toward and Anthropology Beyond the Human*, Berkeley: University of California Press, 71–103.

Landecker, H. (2013) Postindustrial metabolism: Fat knowledge, *Public Culture*, 25(3 (71), 495–522.

Levitt, T. (2020) KFC admits a third of its chickens suffer painful inflammation, *The Guardian*. Available at: https://www.theguardian.com/environment/2020/ jul/30/kfc-admits-a-third-of-its-chickens-suffer-painful-inflammation

Libeskind, D. (1991) *Between the Lines*. Available at: https://libeskind.com/publishing/between-the-lines/ accessed 02/04/2019

MacCormack, P. (2012) *Posthuman Ethics: Embodiment and Cultural Theory*, London and New York: Routledge.

McDonald, B. (2000) "Once you know something, you can't not know it" an empirical look at becoming vegan, *Society and Animals*, 8(1), 1–23.

Mol, A. (2008) I eat an apple. On theorizing subjectivities, *Subjectivity*, 22(1), 28–37.

Nancy, J.L. (1990 [1990]) Finite history, in ed. D. Carroll, *The States of "Theory": History, Art, and Critical Discourse*, Stanford, California: Stanford University Press, 149–172.

Oliver, C. (2020) Beyond-human worlds: Negotiating silence, anger & failure in multispecies research. *Emotion, Space and Society*, https://doi.org/10.1016/j.emospa.2020.100686

Povinelli, E. (2018) Horizons and frontiers, late liberal territoriality, and toxic habitats, *E-Flux*, 90. Available at: https://www.e-flux.com/journal/90/191186/horizons-and-frontiers-late-liberal-territoriality-and-toxic-habitats/ accessed 26/11/2019

Rothfels, N. (2002) *Representing Animals*, Indiana: Indiana University Press.

Scarry, E. (1985) *The Body in Pain: The Making and Unmaking of the World*, Oxford: Oxford University Press.

Serres, M. (1982) Theory of the quasi-object. *The Parasite*, Minneapolis, MN: University of Minnesota Press, 224–234.

Shaw, I.G.R. (2010) Sites, truths and the logics of worlds: Alain Badiou and human geography, *Transactions of the Institute of British Geographers*, 35(3), 431–442.

Simmel, G. (2008) The stranger. in eds. T. Oakes, and P.L. Price, *The Cultural Geography Reader*, Routledge, 323–327.

Smith, P. and Daniel, C. (1975) *The Chicken Book*, Georgia: University of Georgia Press.

Spannring, R. (2019) Ecological citizenship education and the consumption of animal subjectivity, *Education Sciences*, 9(1), 41.

Taylor, S. (2011) Disability rights and animal studies, *Qui Parle*, 19(2), 191–222.

Townley, C. (2010) Animals as friends, *Between the Species*, X, 45–59.

Tsing, A. (2015) *The Mushroom at the End of the World: On the Possibility of Life in Capitalist Ruins*, Princeton and Oxford: Princeton University Press.

Wadiwel, D. (2018) Chicken harvesting machines: Animal labor, resistance, and the time of production, *The South Atlantic Quarterly*, 117(3), 527–549.

Walker, A. (2013) *The Chicken Chronicles*, London: Weidenfeld and Nicolson.

Weaver, H. (2013) Becoming in kind: Race, class, gender, and nation in cultures of dog rescue and dogfighting, *American Quarterly*, 65(3), 689–709.

Willett, C. (2014) *Interspecies Ethics*, New York: Columbia University Press.

Chapter Seven Bibliography

Alaimo, S. (2012) States of suspension: Trans-corporeality at sea, *ISLE: Interdisciplinary Studies in Literature and Environment*, 19(3), 476–493.

Barad, K. (2007) *Meeting the Universe Halfway: Quantum Physics and the Entanglement of Matter and Meaning*, Durham and London: Duke University Press.

Bennett, C.E., Thomas, R., Williams, M., Zalasiewicz, J., Edgeworth, M., Miller, H., Coles, B., Foster, A., Burton, E.J. and Marume, U. (2018) The broiler chicken as a signal of a human reconfigured biosphere, *Royal Society Open Science*, 5(12), 180325.

Blecha, J. and Leitner, H. (2014) Reimagining the food system, the economy, and urban life: New urban chicken-keepers in US cities, *Urban Geography*, 35(1), 86–108.

Borkfelt, S. (2011) What's in a name?: Consequences of naming Non-human animals, *Animals*, 1(1), 116–125.

Braidotti, R. (2018) A theoretical framework for the critical posthumanities, *Theory, Culture & Society*, 36(6), 31–61.

Britton Clouse, M. (2015) *Chicken Run Rescue*. Available at: http://www.chickenrunrescue.org/Municipal-Regulation

Castree, N. (2000 [2000]) The production of nature, in eds. E. Sheppard and T.J. Barnes, *A Companion to Economic Geography*, Oxford: Blackwell Publishing, 275–289.

CIWF (2012) The life of laying hens, *CIWF*. Available at: https://www.ciwf.org.uk/media/5235024/The-life-of-laying-hens.pdf

Davis, K. (2009) *Prisoned Chickens, Poisoned Eggs*, Tennessee: Book Publishing Company.

Druce, C. (2013) *Chickens' Lib*. Bluemoose Books.

Federici, S. (2020) *Beyond the Periphery of the Skin*. PM Press.

Freeden, M. (2015) Silence in political theory: A conceptual predicament, *Journal of Political Ideologies*, 20(1), 1–9.

Godlovitch, R. (1971) Animals and morals, *Philosophy*, 46(175), 23–33.

Greif, M. (2016) *Against Everything*, London and New York: Verso Books.

Hobson, K. (2007) Political animals? On animals as subjects in an enlarged political geography, *Political Geography*, 26(3), 250–267.

hooks, bell (2000) *All About Love: New Visions*, New York: HarperCollins.

Hwang, S. (2013) *The Hen Who Dreamed She Could Fly*. Westminster: Penguin Books.

Jordan, J. (2001) Why Friends Shouldn't let friends be eaten: An argument for vegetarianism, *Social Theory and Practice*, 27(2), 309–322.

Kohn, E. (2013) The living thought. *How Forests Think: Toward and Anthropology Beyond the Human*, Berkeley: University of California Press, 71–103.

Kondaiah, N. and Panda, B. (1992) Processing and utilization of spent hens, *World's Poultry Science Journal*, 48(3), 255–268.

Kristensen, H.H., Berry, P.S. and Tinker, D.B. (2001) Depopulation systems for spent hens—A preliminary evaluation in the United Kingdom, *Journal of Applied Poultry Research*, 10(2), 172–177.

Kundera, M. (1992) *Immortality*, New York: HarperCollins Publishers.

Lorimer, J. (2020) *The Probiotic Planet: Using Life to Manage Life*. Minneapolis, MN: University of Minnesota Press.

Lorimer, J. (2012) Multinatural geographies for the anthropocene, *Progress in Human Geography*, 36(5), 593–612.

MacCormack, P. (2014) *The Animal Catalyst: Towards Ahuman Theory*, London: Bloomsbury Academic.

Mellen, S. (2020) Chicken rehoming charity gets 52,000 lockdown hen requests, *BBC*. Available at: https://www.bbc.co.uk/news/uk-england-53832858

Midgley, M. (1983) *Animals and Why They Matter*, Georgia: University of Georgia Press.

Morin, K. (2018) *Carceral Space, Prisoners and Animals*, Oxon and New York: Routledge.

. Nancy, J.L. (1990 [1990]) Finite history, in ed. D. Carroll, *The States of "Theory": History, Art, and Critical Discourse*, Stanford, California: Stanford University Press, 149–172.

Nancy, J.L. (2014) The political and/or politics, *Oxford Literary Review*, 36(1), 5–17.

Newberry, R.C., Webster, A.B., Lewis, N.J. and Van Arnam, C. (1999) Management of spent hens, *Journal of Applied Animal Welfare Science*, 2(1), 13–29.

Rich, A. (2003 [1984]) Notes towards a politics of location, in eds. R. Lewis and S. Mills, *Feminist Postcolonial Theory: A Reader*, New York: Routledge, 29–42.

Scarry, E. (1985) *The Body in Pain: The Making and Unmaking of the World*, Oxford: Oxford University Press.

Searle, A. and Turnbull, J. (2020) Resurgent natures? More-than-human perspectives on COVID-19. *Dialogues in Human Geography*, 2043820620933859.

Smith, P. and Daniel, C. (1975) *The Chicken Book*, Georgia: University of Georgia Press.

Smith, S.J. (1999) The cultural politics of difference, in eds. D. Massey, J. Allen and P. Sarre, *Human Geography Today*, Cambridge: Polity Press.

Stanescu, J. (2012) Species trouble: Judith Butler, mourning, and the precarious lives of animals, *Hypatia*, 27(3), 567–582.

Vermeulen, P. (2017) "The sea, not the ocean": Anthropocene fiction and the memory of (Non)human life, *Genre: Forms of Discourse and Culture*, 50(2), 181–200.

Vestergaard, K. (1987) Alternative farm housing: Ethological considerations, *Scientists Center Newsletter*, 9(3), 10–11.

Wadiwel, D. (2016) Fish and pain: The politics of doubt, *Animal Sentience*, 3(31), 1–8.

Wadiwel, D. (2018) Chicken harvesting machine: Animal labor, resistance, and the time of production, *South Atlantic Quarterly*, 117(3), 527–549.

Weaver, H. (2013) Becoming in kind: Race, class, gender, and nation in cultures of dog rescue and dogfighting, *American Quarterly*, 65(3), 689–709.

Wolch, J. and Emel, J. (1995) *Animal Geographies: Place, Politics, and Identity in the Nature-Culture Borderlands*, London: Verso Books.

Conclusion Bibliography

Adams, C.J. and Gruen, L. (2014) *Ecofeminism*, London: Bloomsbury.

Cochrane, A. (2018) *Sentientist Politics: A Theory of Global Inter-Species Justice*, Oxford: Oxford University Press.

Davis, A. (2016) *Freedom Is a Constant Struggle: Ferguson, Palestine, and the Foundations of a Movement*, Chicago, Illinois: Haymarket Books.

Derrida, J. (2005) *The Politics of Friendship*, New York: Verso Books.

Donaldson, S. and Kymlicka, W. (2015) Farmed animal sanctuaries: The heart of the movement, *Politics and Animals*, 1(1), 50–74.

Gillespie, K.A. (2019) For a politicized multispecies ethnography. *Politics and Animals*, 5, 17–32.

Greif, M. (2016) *Against Everything*, London and New York: Verso Books.

Hinchcliffe, S. and Whatmore, S. (2006) Living cities: Towards a politics of conviviality, *Science and Culture*, 15(2), 123–138.

hooks, bell (2000) *All About Love: New Visions*, New York: HarperCollins Publishers.

McLemore, S.D. (1970) Simmel's 'stranger': A critique of the concept, *Pacific Sociological Review*, 13(2), 86–94.

Nancy, J.L. (1990 [1990]) Finite history, in ed. D. Carroll, *The States of "Theory": History, Art, and Critical Discourse*, Stanford, California: Stanford University Press, 149–172.

Nancy, J.L. (2014) The political and/or politics, *Oxford Literary Review*, 36(1), 5–17.

Nelson, S. and Braun, B. (2017) Autonomia in the anthropocene: New challenges to radical politics, *South Atlantic Quarterly*, 116(2), 223–235.

Scott, D. (1999) *Refashioning Futures: Criticism after Postcoloniality*, New Jersey: Princeton University Press.

Spannring, R. (2019) Ecological citizenship education and the consumption of animal subjectivity, *Education Sciences*, 9(1), 41.

Todd, Z. (2015) Indigenizing the anthropocene. *Art in the Anthropocene: Encounters among Aesthetics, Politics, Environments and Epistemologies*, 241–254.

Wrenn, C.L. and Johnson, R. (2013) A critique of single-issue campaigning and the importance of comprehensive abolitionist vegan advocacy, *Food, Culture & Society*, 16(4), 651–668.

Index

Printed in the United States
by Baker & Taylor Publisher Services

Printed in the United States
by Baker & Taylor Publisher Services